# BACKYARD SHEEP FARMING

# backyard sheep farming

by

## ann williams

Published in Great Britain 1978 by

PRISM PRESS

Copyright Ann Williams 1978
Reprinted 1988

SBN 0 904727 71 8 Hardback Edition

SBN 0 904727 72 6 Paperback Edition


Printed by

A. Wheaton & Co. Ltd, Exeter

# contents

# acknowledgements

Many people have helped with this book and I owe thanks to them all. In particular I would like to thank Hazel Watson, of the Ripon Library, for help in obtaining books. Brian Sidgwick and his daughter Irene have given me the benefit of their experience with backyard sheep and also many contacts in the sheep world. Harry Topham told me how a sheepskin should be cured and Malcolm MacDougall gave advice on spinning and weaving. Leonard Chandler and his wife read the sheep section and Peter Clarke, MRCVS made forthright and helpful comments on the manuscript.

Then I thank my family, because our backyarding has been a family affair; my parents, who embarked on a backyard adventure which made all our lives more interesting and my brothers, David for ideas and suggestions and Peter for his help with research.

Thanks are due to the Soil Association for allowing me to quote from their article on sheep dips and to reproduce the list of approved dipping products. I would also like to thank the Jacob Sheep Society for their permission to quote from Lady Aldington's notes for beginners.

Finally I would like to set down my appreciation of the encouragement given me by Colin Spooner, the publisher of this series of backyard books.

Ann Williams

# 1 what about sheep?

*Backyard Sheep* may sound like a non-starter, but I was surprised to find when I started to think about this book that there are many families with a small paddock or orchard, who keep a small flock of sheep. In some ways they are easier to look after than pigs.

Sheep will hardly be happy in a concrete yard, although more and more commercial flockmasters are wintering their sheep indoors. There are breeds, though, that will be perfectly content with a small field provided that it can be given a rest from time

to time. Other breeds of sheep are unsuitable for backyarding; the Welsh Mountain breed will not settle down to anything less than a Welsh mountain.

Backyard sheep are there if you look for them. In North Wales every other council house tenant is a part-time shepherd, since in some villages everybody has grazing rights on common land. In villages in Yorkshire, most houses lining the street have a paddock behind the house where you will see little flocks of docile family pets. Some small flocks are kept just to eat the orchard grass, with the added bonus of the wool.

Why keep sheep? Because once you get to know them, you will like them. Sheep are amiable and quite intelligent. The danger may be (as with any kind of animal) that once you get interested in them, your conversation will tend to stray back to the one subject; this being the case with shepherds, you can learn a lot from them.

Here are some more practical reasons for keeping sheep:

They are efficient producers from land which would not support cattle; in fact sheep do rather better on the poorer sort of land, provided that they get enough room. They can do very well on rough grazing, but not so well if the land is wet. Good slopes are what they like, for exercise, shelter and for drainage. Three or more sheep can be kept where one cow would graze on ordinary grassland.

Sheep produce meat; say two lambs a year per ewe, which can be cut up for the freezer to give you, at a rough estimate, about 70 to 80 lbs of lamb. The prices of meat vary so much that I will not attempt to quote them, but lamb compares well with pork whatever the price is doing. The sheep eats more roughage and needs much less concentrate food. Lambs can be fattened on good grass and little else, which must be a good way of using a small paddock.

Wool is a very nice thing to have. Once you possess a fresh glossy fleece of your own to do with as you wish, a whole new world opens up. There are fascinating crafts to be learned - spinning, weaving and dyeing the yarn with plant dyes from the hedgerows  Many people are now showing interest in these old crafts and there are opportunities (which did not exist a few years ago) to learn how to do these things. It is now quite easy to obtain the tools and the advice that you need.

I said that some small flocks are kept to eat down the orchard grass; the docile breeds are good lawn mowers. Some can be tethered during the day and they will keep the grass nibbled down which can be very useful.

The fertility which a sheep spreads so evenly, treading it in with dainty feet, used to be all-important. It could be so again. In this country we are dependent on imported chemicals for much of our artificial fertiliser  and if the system eventually breaks down, or if people begin to recognise that chemicals are not all that the land needs, then the sheep may come into its own again in the lowlands. For backyard fertility sheep are excellent - less messy than cattle, not as smelly as pigs.

Sheep are amazingly good at recycling garden waste. As ruminants (see The Backyard Dairy Book for details) they are well equipped to deal with bulky, fibrous food and they appreciate

11

a change of diet. If you have a large garden they could help you clear up both weeds and waste.

If you are already backyarding and have other classes of stock, the sheep will fit in well with them. I once saw a series of trial plots at a Ministry of Agriculture experimental farm, where they were trying to find out the benefits of mixed grazing. It was obvious that when sheep and cattle grazed the same land. either together or in turn, the land and the animals did better than with cattle or sheep alone. Apparently a particularly good combination is horses and sheep. This advantage comes from several reasons; types of grasses favoured by different species and also parasites differ between species allowing, in the case of sheep and horses, an opportunity to interrupt the life-cycle of each other's worms.

If you are, to quote the Backyard Dairy Book again, desperate for the taste of backyard Roquefort, you can milk sheep as they do in France to produce this famous cheese. For centuries sheep were kept in Britain mainly for milk and wool. A man I know has made sheep's milk cheese when he had a few ewes with a lot of milk and no lambs to feed. He said that the cheese was delicious and very rich.

## A Bit of History

Our quiet woolly sheep are rather different from their wild ancestors. It seems strange that we cut the tails of our sheep to protect them from becoming dirty and then being attacked by blowfly, while the wild sheep had short tails which took care of that problem. The surviving primitive breeds still have short tails. Presumably man decided to breed them long - but why? The only reason I can think of is that perhaps the tail was a delicacy at one time, like the fat-tailed sheep of Syria.

Sheep are also woollier the further they get away from their wild ancestors. The first sheep will have been hairy little things, rather like goats. Their colouring would also have resembled that of goats. The more domesticated the sheep has become, the more of a little white sheep it is and the less of a black sheep.

At some stage in their evolution, sheep have been persuaded not to moult, so that we can shear them in one piece and not run around after them picking up discarded fleece. Cattle shed their winter coats in spring and acquire a sleek, thin summer coat. It would be natural for sheep to do the same, but very obligingly they keep it on until we relieve them of it.

The story of sheep is as old as the story of mankind. On temple friezes dating back three thousand years B.C., there are sheep skipping about or being milked from behind, which must have been the practice then.

Britain is sometimes called a sheep museum - we have nearly forty different breeds. There have been many changes down the centuries and each successive wave of intruders to these islands has brought with it another type of sheep. This might account for some of the different breeds, but the real reason for them seems to be the fact that the sheep is an adaptable animal, and we have varying conditions in Britain to be adapted to.

First of all there were the small primitive breeds and it is thought that traces of these survive in Scotland, in the old Highland sheep, now nearly extinct, and in the Soay sheep

which have lived on their uninhabited island in the St. Kilda group, and thus kept their blood pure. They still have short tails and brown fleece like wild sheep. There is a similar St. Kilda breed on the main island. In Wales, the mountain sheep have their origins in the remote past, although they have become more evolved. They have long tails and are mostly white, but some of them have traces of brown and they are small sheep. Conditions on the mountains are such that attempts

to cross them with other breeds have not met with much success, so that they are reasonably pure.

It would be fascinating if we could take a piece of mountain land in Western Britain and find out by some magical means just how far back one could trace the ancestors of the sheep that graze the land today. In those parts of the world, a farm is sold with its flocks, which are *hoofed* to that particular hill. Through inheritance the sheep are at home on that piece of land; they know its limits and they will not stray. If taken away from it permanently, they would be lost. When sent away to lowland farms for wintering, they tend to break out and cause trouble.

This instinctive knowledge of the farm limits means that they will stick to their own piece of ground on the open hill, even when the division between one farm and the next is not fenced. It is called *cynhefin* in Welsh, *hoofed* or *hefted* in English; and it has been handed on for hundreds of generations to the lambs born on that hill. I met a man in Wales whose family had lived on the same farm for 600 years, with as far as he knew the same flock of sheep. Usually the men come and go, but the flock goes on. Only a catastrophe, such as a terribly hard winter or a plague like foot-and-mouth, can break the tradition.

The Romans brought sheep to Britain - they thought that the native ones were a bit scruffy - and they started an export trade of wool to Italy that one could almost say has gone on ever since.

The Dark Ages are obscure but when the monasteries were established sheep became very important. Many of the monastic houses were run by Cistercians and others with a European tradition so that ideas from other places filtered through to the remote countryside. As part of their self-sufficient way of life, the monks built up farms, and they led the way with new methods and in stock improvement. This is true of my own area of Yorkshire; we live now between the two great abbeys of Fountains and Jervaulx, and traces of the monks' stay are still evident in the countryside. The monks of Jervaulx are credited with the development of Wensleydale cheese. This was made originally from the milk of Wensleydale ewes, grazing the sweet limestone pastures of the river Ure. Only later was it made from cows' milk.

For a thousand years the sheep was the most important animal in Britain. In a communal flock the villagers of the Middle Ages sent out their few sheep with everybody else's, to graze the common land during the day. This common was wild, untamed land between one village and the next and only a sheep could make good use of it. At night the sheep were

THE FOLD

brought back onto the stubble or fallow, and *folded* - fenced in small enclosures - so that their manure would enrich the land. Fertility was painfully difficult to acquire and thus the sheep was important to the next crop. Meanwhile, it produced food and clothing. The villagers spun their own wool and took

it to be woven. They could also make their own cheese and butter. A family too poor to own a cow could have a sheep or two and this would mean that they could have dairy produce.

This was on the simple level; but nationally sheep were important as well. Much of the prosperity of Britain was built upon the export of wool; without sheep, our history might have been different.

Folding sheep on arable land continued right up to the 1930's when this system of manuring was replaced by artificial fertilisers. Without the *golden hoof* the barley crops, especially on light land, would have been very poor indeed. Sheep were often kept not for their own profit, but for the good they did to the subsequent crops.

For about two hundred years now, our sheep farms have bred for meat as well as for wool. This started when the Industrial Revolution created towns full of people who produced none of their own food; before that, everybody could be a backyarder.

In some cases, the wool suffered when they started to breed for meat. The pioneer Robert Bakewell was famous for his improvements to Leicester sheep; his were called the new Leicester. They were marvellous for meat ( previously sheep had all been a bit scraggy) but their coats were so light that breeders made little jackets to keep the rams warm!

LEICESTER EWE

By the end of the eighteenth century, some of the big owners had taken an interest in improving their livestock and they tried to give a lead to the countryside in better farming. There arose a tradition of *sheep-shearings* which were like open days, when the public could come in and inspect the leading farmers' stock and crops and perhaps pick up a few new ideas. In East Anglia, Mr Coke of Holkham (later Earl of Leicester) was famous for his shearings. They were called 'Coke's Clippings' and they spread ideas about for over twenty years. They had a lot of fun, washing and shearing the sheep, with free food and drink. There were exhibitions of new machinery and competitions for guessing the weight of animals.

The Ministry of Agriculture has tried with varying success to take over as leader in improving farming methods. In some cases they have gone into agribusiness, but if they have an open day near you, go along. The latest findings will be available and then you can choose what seems good among the new ideas. Once you have seen what's new, make up your own mind.

# 2 which sort of sheep?

With nearly forty different breeds to choose from, and about eighty crosses, we should have plenty of choice of sheep. So what kind are best for backyarding?

The short answer is to stick to your local breeds or cross. It will have become adapted to conditions in your area, wherever you live. Backyarding, being a state of mind, can be carried on at a thousand feet in the wet, wild west, or in a Surrey suburb. Get out into your nearest bit of country and find out what sheep the locals favour. Make friends with the shepherds and carry on from there.

Let us consider the main types of sheep and the sort of conditions they prefer.

Firstly, the mountain and hill breeds. One is the Welsh Mountain, descended from the little Celtic sheep which must have come here before the Romans. The Scottish Blackface which can live on heather is another. In Yorkshire we have the Swaledale. All the mountain breeds are very good to eat; as Peacock said, 'The mountain sheep are sweeter but the valley sheep are fatter.' They are all smallish, hardy sheep and can stand a lot of cold and wet. However, they will not do for a small patch of ground and so we cannot really call them suitable for backyarding, although if you happen to have grazing rights on some moorland, a small flock would be possible.

Mountain sheep are most interesting; keeping sheep on the

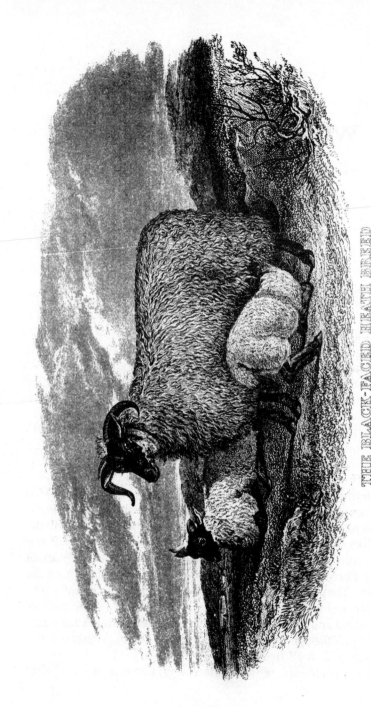

### THE BLACK-FACED HEATH BREED

One Year Old

Bred by Mr Thomas Robertson, Broomlea, County of Peebles.

hills is a way of life and a tough one, even in these days. I heard
a shepherd say recently that he walks twenty miles a day
in the lambing season.

If you live in an upland area you may well get interested in
the local sheep. To keep a few sheep on the moor, it will be
absolutely essential to use the local breed; you will have to buy
some from a neighbour and be guided by his advice for quite
a long time in order to get sheep which have been born on and
acclimatised to that moor. You may think that the locals are
a bit hidebound, but tradition is everything with hill sheep.
There is a certain date for all the different jobs, rather like a
ritual, but these have been found over hundreds of years to be
right and you have to fit in to local tradition. It can be very
absorbing and educational, and in moorland areas you find
sheep enthusiasts who will enjoy telling you all about them.

True moorland sheep are only kept for breeding for three
or four seasons. They need to be young and to have all their
teeth to do well in rugged conditions. After this, they are sold
at the autumn draft sales to farmers lower down the hill, where
the middleaged ewes will have an easier time. They are then
crossed with a Down ram to produce fine hardy lambs. The hill
men also export their mountain lambs for fattening on the lowlands.

Ewe lambs of the mountain breeds, which are kept for breeding,
are often sent away for their first winter to farms with easier
conditions so that they continue growing and make better ewes.

21

EWE & LAMB, SOUTH DOWN BREED

Bred by Thomas Ellman, Esq.ʳ Beddingham.

W. Tuffen

We used to take in sheep like this for the winter. They improved the grassland, but they were terrors to keep in. A mountain ewe can walk up a stone wall and dance on the top; which is why the wild beautiful scenery in mountain districts is defaced by scrawls of wire netting.

I believe that the primitive breeds of sheep are difficult to shepherd, since they scatter instead of herding together. Welsh sheep must have some primitive blood left as they are not easy to deal with always. I have seen Welshmen stamping on their caps and screaming with rage when trying to gather them. But these were large flocks which had most of Snowdon to roam on, so no wonder they were wild. Your backyard sheep will never behave in this way since they will be much more used to human contact.

The short-woolled breeds are the next type of sheep to consider. These are mainly lowland sheep, the sort that used to be folded in very small pens to improve the land. They are therefore docile and easy to handle and consequently make good backyard sheep. Often they are compact, rounded white sheep with either black faces and feet like the Hampshire and Oxford Down, or a white face like the Southdown. They have, as the name suggests, short curly wool.

On the lowlands sheep are kept now on grassland for the production of fat lambs, and sometimes on arable farms to provide a 'break' or change from corn. Some are kept inside in large sheds for the winter to keep them off the land when it is wet. In-wintering would not go down well with the mountain breeds, but these lowland sheep seem not to mind it. They tend to live in flocks, staying in closeknit groups, whereas mountain sheep have to scatter to find their food.

The long-woolled breeds have rather gone out of fashion in their pure state because they are big sheep and their meat is coarse, with large joints. They also produce a lot of coarse wool, a fleece will weigh about 15lbs. These breeds are still used extensively for crossing with other breeds. For instance, the Border Leicester, that distinctive Roman-nosed sheep which evolved from Mr. Bakewell's New Leicesters, is crossed with the Cheviot (a hill breed) to make the Halfbred, which is a very popular commercial cross and has almost reached the status of

LEICESTER RAM

Bred by and the Property of M<sup>r</sup> Dickinson. Magdalene Hall, Roxburghshire.

24

CHEVIOT EWE

Bred by and the Property of Mr Thomas Elliot, Hyndhope, Roxburghshire

25

a breed.

There are several important crosses, for which the long-wool breeds are in demand:

Border Leicester  x  Scottish Blackface = the Mule
Wensleydale  x  Blackface or Swaledale = the Masham
Border Leicester  x  Welsh Mountain = the Welsh Half-bred

So now we come to the Jacob's sheep which is possibly the best breed for back-yarding, for reasons which will become apparent. They are nearly always seen in small flocks. If you see a distinctive spotted sheep, a creamy colour with chocolate spots and two, four or six horns, that is a Jacob.

This is a very old breed with a fascinating history. Their story is told in the Book of Genesis — you may remember it. When Jacob went to work as a shepherd for Laban; he was there seven years, and asked for Laban's daughter to be his wife. Instead he was given her elder sister Leah; he had to work another seven years for Rachel.

So he had done fourteen years of shepherding and when he had got Rachel, he decided to go home. He had not been given any wages in all these years, so he asked Laban, and was promised, all the spotted sheep and goats from the herds as a reward for his years of toil.

Then Laban cheated. He gave the spotted animals to his own sons and there were none left for Jacob; so the shepherd laid his plans.

He took rods of willow, hazel and chestnut and peeled off the bark in patches to give a spotted effect. These he stuck in the ground near the well to which the flocks came to drink. It must have been the breeding season; all the lambs and kids which had been conceived in the sight of these rods turned out spotted. Jacob waited for the offspring to be born, and then he went off home with his flock, all of them spotted. Jacob's sheep have been spotted ever since. This theory of breeding is often quoted in the old books, with a warning that one should be careful about what one's pedigree stock is looking at, at the moment of conception. The story was quoted in many a breeding manual, even after Darwin and Mendel.

In spite of Jacob's story, students of breed history think that the Jacob may be related to either the multi-horned Hebridean sheep or those of the Mediterranean.

Jacob sheep are very attractive and they were first imported purely for ornament, to inhabit parkland as a sort of novelty. Being hardy, they have thrived here but they were getting scarce when in 1969 the Jacob Sheep Association was formed to promote their interests. Now, they are popular for backyarding. If you have no contacts but would like to see some Jacob sheep, the Society would put you in touch with breeders, as would the National Sheep Association for any of the other breeds. Jacob breeders seem to be real enthusiasts. The sheep add to their advantages of hardiness and docility a wool which is suitable for hand spinning and good meat - small, sweet joints with little fat. If you want a larger lamb, the Jacob crosses well with a Down ram.

If you are trying to choose a breed or cross, you may want one that is a useful wool producer, so that you can try spinning. There is so much variety in sheep's wool that perhaps we should digress for a moment and consider wool.

The main variation in fleeces depends upon the breed, but there are other factors. Examples are the climate and the sheep's food. If it is short of food or in poor health, the wool will be poorer, and this sometimes happens just for a short time so that there is then a weaker band in the wool, where the set-back has occurred.

The length of the fleece can vary from about 2 to about 15 inches. The longwools, as their name implies, produce the longer fleeces. The weight of the fleece when sheared can be between 2½lb ( for a Welsh Mountain) and about 15lb ( Border Leicester).

Wool fibres come in locks called *staple*. They may be curled and/or crimped; curl is the bend along the length and crimp is the wave across it. These also vary according to breed.

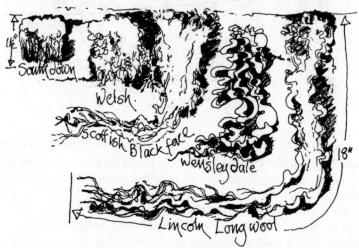

The fibres are also found in varying degrees of fineness. This is assessed by a count which is the number of hanks (560 yards) of yarn which can be spun from one pound of wool; the finer the yarn, the greater the length for the same weight of course. Merino wool is the finest and can get up to a count of almost 100. British wool ranges from 56-60 for down wool to about 28 in some mountain fleeces.

Merino or Short-wool Fleece, for Woollens.

There is also hair on sheep, known as kemp. Some breeds have more than others; there tends to be more on sheep which have adapted to wet climates because it sheds the rain. Kemp is noticeable on rough tweeds as the hairy bits that stick out of the cloth and have not taken the dye.

The oiliness of a natural fleece is obvious as soon as you handle it. This weatherproofs the coat for the sheep and if the oiled wool is knitted up without washing, as in the fishermen's sweater, the weatherproofing remains. The oil is produced like that in human hair; each fibre has a little oil gland at the base to lubricate it. For spinning by hand the fleece is not usually

29

washed, unless you want to dye it, because it is easier to spin when oily. The effect of the oil wears off a little when the fleece is stored, and sometimes olive oil is used to soften the wool again for hand spinning. There is another, sweaty substance in the fleece called *suint* which adds to the greasiness of a fresh fleece.

*Lustre* is another term applied to wool. You may have noticed that some wool seems to have a shine about it, while another sort may look dull. This depends on the scales of each fibre. When the scales are large and flat the wool is smoother and has more lustre. Long wool is often more shiny than short; the different lustres are used for different types of cloth in the wool trade.

A further complication about wool is that the quality varies on different parts of the fleece. It has therefore to be sorted into piles of the various qualities, as a preliminary to spinning. The wool on the shoulder is the best quality.

Of course the life the sheep is leading will be reflected faithfully in the fleece. Apart from the quality differences due to health and quantity of food, even the surroundings of the sheep have an effect. If it has been creeping about in undergrowth, the fleece will have picked up burrs, seeds and small twigs. Sometimes rubbing against a trough picks up wood splinters in the fleece. The weather will affect the fleece - sunshine will bleach the back.

If you feel rather nervous about facing a whole fleece for your first efforts at spinning, you can always go round the field (or out on the moors) and pick up all the little pieces of wool caught on bushes and fences. In this way you can gain confidence in handling wool.

## Milk Sheep

Since this book first appeared, backyarders in Britain have begun to take an interest in continental breeds of sheep which might be suitable for milk production. This tradition has been kept up, notably in the Roquefort area and in Italy, and a triple purpose sheep certainly sounds attractive. At the time of writing they are scarce in Britain but perhaps there will be more available as interest grows. My fear would be that high yielding sheep from Europe might not be hardy enough for cold backyard winters, but they could be worth a little coddling.

The East Friesland seems to be the best for milk production, and oddly enough it comes from the same part of the world as the Friesian cow. There is said to be a dash of Leicester blood in this sheep's ancestry.

It seems you can expect a high performance from a Friesland ewe, but she will need a lot of food to keep her going. On average they produce two lambs a year, about 200 gallons of milk and a fleece weighing 10 to 12lb.

You should be able to recognise a Friesland sheep because they have a distinctive tail, thin and bald like a rat's tail. Lawrence Alderson says in his Observer's Book of Farm Animals that the main function of the East Friesland in Britain is to improve the native longwool breed; but he does say it has been known as the 'smallholder's cow'.

# THE VOCABULARY OF SHEPHERDS

Sheep names vary according to district. Even the word 'lamb' is rather vague, although strictly it belongs to an ovine animal that has not got its first pair of permanent incisors.

## Females

| | |
|---|---|
| Ewe lambs, gimmer or chilver | — birth to weaning |
| Ewe teg, gimmer hog, ewe hog | — weaning to first shearing |
| Shearing ewe, shearing gimmer gimmer or theave | — first to second shearing |
| Ewe | — has had a least one lamb |
| Crone | — an old barren ewe |

## Males

| | |
|---|---|
| Ram lambs, hoggets, hogs | — weaning to first shearing |
| Shearing tups or rams | — after first shearing |
| 2, 3, 4 shear rams or tups | — according to number of times they have been shorn |

## Castrated Males

| | |
|---|---|
| Wether or wedder | — weaning to first shearing |
| Shearing wether | — first to second shearing |

# 3 starting with sheep

Having decided what kind of sheep to keep, how do we start?
Beginning with pet lambs is a slow way, but in some ways a
good one. These are orphan lambs which have either been ab-
andoned by their mothers or lost their mothers through death
or accident. Shepherds of large flocks always have one or two
at lambing time and they rarely have the time to look after
more than a few themselves, so they may be glad to give, sell
or swap them to a backyarder.

A backyard flock I know well was started in this way. It
is easy to find motherless lambs in this district because we are
on the edge of the moors and among sheep of all breeds.Some
of the lambs which my friends started with cost them £1, and
some they exchanged for services rendered. They took male
and female lambs and reared them all the first year on the
bottle, using a proper lamb teat which you can buy easily
for this job. They mixed up a ewe-milk substitute (Ewelac)
for the lambs; like the artificial pig milk it is expensive, but
reliable. Later, they found that goat milk was cheaper.

Lambs reared on the bottle are very time-consuming at
first. They have to be fed every few hours, but if you get up
in the night for one, you may as well feed half a dozen! Weakly
lambs will need a bottle in the night, but with healthy or older
lambs you can get out of it. Give them a late-night feed and an
early morning one and they will be happy enough. But the job

of rearing lambs on the bottle is only really feasible if there is someone at home all day; if you all go out to work, it would be difficult to give them enough attention. The instructions for feeding are given with the milk but the routine is a matter of common sense; bottles and teats must be kept very clean and any lamb which shows signs of scouring must be taken off milk and given glucose and warm water. Since they are greedy little things, this is sometimes necessary.

The milk of a ewe is fairly rich - about 7% fat (cow's milk averages about 3.5% fat) and any substitute for sheep's milk therefore needs to be rich. If the kind of milk powder sold for calves is all you can get, mix it with rather less than the recommended amount of water. Cow's milk is quite good for older lambs, and they like goat's milk.

Lambs reared on the bottle get rather rough as they grow and of course they are very tame. My friends got used to

hungry lambs trotting into the kitchen to look for the bottle. This led to full lambs sitting by the fire, which led in turn to half-grown sheep wanting to sit on people's knees. No doubt a line has to be drawn somewhere, and it might be as well to decide quite early just where that line should be, before the sheep get a chance to take over. The great advantage of these tame sheep is that no dog is needed to round up the flock; as soon as you appear in the field they will come to you.

This was a slow way of starting a backyard flock, but it did work. When fat, the male lambs were sold to the butcher and this paid for the milk substitute and the cost of the lambs. The ewes were put to the ram at about 1½ years old, at the end of their second summer. They are now mature sheep with lambs of their own, but they still walk up to people instead of away. They all have their names and distinct personalities, they even have different voices. This is just as it should be to a good shepherd, of course, and when you rear them yourself there is no danger of not knowing each one personally. This is shepherding made easy - you learn without thinking about it.

So far, my friends have not regretted their venture and the hand-reared ewes have made good mothers. But perhaps I ought to mention that some people believe that pet lambs make bad mothers. This could be merely an inherited characteristic; they had perhaps been rejected by their own mothers, which is why they came to be pet lambs! All I can record is that the lambs I know personally have made a good job of rearing their own lambs.

You can train sheep to do all kinds of things, like dogs, when they are tame. They learn by repetition, but they have better memories than dogs. However, you cannot house-train a sheep. Unlike pigs, which are born in a nest, sheep are born in the open field. It does not matter where the droppings go, so they have no instinct at all to base any training on.

As I said before, lambs like goat's milk, and if goats are one of your backyarding activities this will be a godsend when it comes to rearing lambs. The safest way will probably be to milk your goats in the normal way and then feed the milk in a bottle to the lambs; but you can rear lambs by direct goat power. A goat in full milk will rear three or four lambs, and I heard of one doing seven this spring! A goat may not however allow the lambs to help themselves. They have to be kept in a warm pen and plugged into the goat at frequent intervals, two at a time. Lambs drink very quickly - watch how soon they are full when drinking from the ewe - so they do not need to be left on the goat for very long. Their stomachs should change from slightly hollow to straight sided; if they get round, they have had too much. A lamb can soon get indigestion or 'blow up' with wind and the best thing to do is to watch closely to prevent this happening. If it does, dose with milk of magnesia.

Rearing lambs is a slow way to start but it could be a good thing because it gives you time to learn about them. If you are impatient and want to go a bit faster then the answer may be to buy draft ewes. You may remember that hill flocks cull their ewes after three or four lamb crops; there are huge autumn sales in all the sheep districts. These sheep may be reasonably sedate (one hopes) because of their age, but they are mountain sheep, after all. They will want a reasonably good range, so do not shut up sheep of this sort on a small piece of ground. But if you have some rough land, you could try a few draft ewes and feed them by hand to tame them, if they will take food. Some hill sheep are not given any food, not even hay, that they they do not find for themselves. They don't know what it is at first.

In the end it might be difficult to establish a relationship with hill ewes. The solution would be to keep their ewe lambs (which could be half lowland blood if you used a Down ram) and get to know them while they are young. Then sell the stand-offish mothers to a larger flock.

Starting the expensive way would mean that you bought ewe lambs at weaning from a small flock; this might be the

best way to start with Jacob sheep because they are not very common and would be hard to find as orphan lambs or draft ewes; in fact as an enthusiast's breed they are not often treated on purely commercial lines.

Sheep are fairly easy to buy and sell. We bought a batch of in-lamb ewes to graze off some very old rough pasture, that we had just taken over. This land needed to be eaten off bare, harrowed over and manured during the winter. This was the alternative to ploughing it up and re-seeding with grass; it was better to use sheep, for several reasons. It was cheaper, and also the old pastures contain a lot of herbs and mixed grasses that grazing animals like better than single-species new leys; it was a pity to plough these out. And also, in this case the field was on a hillside and there was rock near the surface which would have made ploughing difficult. I have digressed a little, but I mention our little flock because sheep can be used in this way as part of grassland management.

When you have ewes ready for breeding, a ram will have to be found. Ewes come into season in the autumn, and at 16-18 day intervals. For a small flock, it is hardly worth keeping a ram of your own. Try to find a farmer who is willing to let your sheep run with his for about six weeks. This covers two heat periods and should ensure lambs for them all. Or perhaps he would lend you one of his spare rams - large flocks have several rams and use them in rotation, just in case one turns out to be sterile. So a 'resting' ram might be boarded out with your sheep as a useful arrangement for all.

Lady Aldington, who is Chairman of the Jacob Sheep Society, has some words of warning to say about rams. 'Never have a 'sock' ram (i.e. one who has been brought up on a bottle). He will end by butting you in affection.... do not keep a ram in a field next to ewes, once the ewes have come into season, in September to October; it is unkind to both; I've known one go mad, butting everything in sight, and it finally had to be put down.'

The breed of ram will depend on what you can beg or borrow; but try for the best you can. For fat lambs, a ram of the Down breeds will be best and they are common, since the farmers are aiming for the same thing.

If the ewe lambs are well-grown, they can run with the ewes and they will then meet the ram and will probably breed. Make sure, though, that they are well enough grown. Lambs reared on the bottle have often had a tough start in life; they may be small and thin when you get them, and perhaps they will not have caught up by the autumn. If this is the case, take them out of the flock before the ram goes in and let them grow, and breed the next year. They will be better for it.

When should we put the ram in with the ewes? The gestation period is five months; so if the ram goes into the field about 5th November, lambing will start in early April, which should be soon enough. Early lambs are most pathetic in a hard winter; in fact April can be very cold sometimes.

As with any livestock, before they arrive the place will have to be got ready for them. A paddock or orchard will want really good boundaries to keep them in. 'Good fences make good neighbours', and this is particularly true with sheep. If they get into your garden and eat off all your cabbage plants this may put you off sheep for years. If your sheep eat your neighbour's garden, it may put him off sheep, and you, for good. What my friends call 'creepers' are the worst; they are bad sheep that make a study of escapology. Sometimes they can be cured of this vice by tethering for a while. (Farmers used to tie their legs together, which is now illegal, or fix some framework around the horns to stop them going through small gaps. Ewes had the frame fixed around the neck) Good post and rail wooden fencing, well creosoted, can be used to supplement gaps in hedges. The hedge will grow round the fence in time.

Barbed wire is horrible stuff; keep away from it if you can. Three strands are needed to keep sheep in, and even then the lambs will be under it. Sheep netting is expensive but good; it will keep in everything, even small lambs. I have noticed that serious lowland sheep farmers make much use of it.

There is now special electric fencing for sheep, a sort of plastic electrified sheep wire. At the moment of writing this costs £17 per roll of 50 metres. It is a complete fence; the posts are fastened on to the wire and I am assured that it will keep in lambs.

For the lawn-mowing job a tether would be best, and of course a nice docile sheep, well used to being tied up. A quiet animal can be got used to tethering gradually, for short periods at first. Make sure that water and shade are within reach if you intend to leave the sheep alone for any length of time; also check that there is nothing lying about for the sheep to get tangled up in.

*Tenting* is an old pastime that could well be revived. It used to be a way for smallholders without much land to stretch the grazing land give their animals a bite of something fresh. This is a Northern expression (probably a corruption of 'tending') for letting your animals graze bits of ground, the sides of country lanes, commons and so on. You stay with them and keep them from straying into fields and gardens. It is a nice peaceful occupation where there is no traffic, and children can do it as they used to. If women did the job, they knitted or spun at the same time. It really amounts to taking your

39

animal for a walk at its own pace. A docile sheep, used to being tethered, will enjoy being 'tented' along a quiet lane.

However we graze the sheep, winter will come and the winds will blow cold. Some people do not believe in giving sheep too much shelter; but I think they will need it, for lambing at least. The same rough shed will do for shade in summer and shelter from the wind in winter. I think that it is more important to provide shelter for backyard sheep than for those which are roaming the range; these have much more choice of where to go to get away from the weather. The sheep shelter need not be elaborate; a poultry house is suitable, or an open-fronted shed with pens rigged up in it. If there is an old barn or stable in the field, this is perfect, if not, straw bales will break the wind - or you can make a straw-bale hut.

It really does help to have your sheep near the house if possible, at lambing time; so if you have a small home paddock, save it for this time of year.

# 4 routine with sheep

The most important routine with sheep is to see them at least once every day of the year. At lambing time or in extreme weather conditions, it should be more often.

A shepherd looks round his sheep and notes what they are doing every time he sees them. Then if something is slightly wrong, it will be noticed right away because subconsciously he has built up a knowledge of their normal behaviour. A good stock man often knows there is something wrong before he has any idea of what it is.

What are we commonly looking for, every day of the year? I am afraid we are looking for trouble. Firstly, any signs of illness - sheep looking miserable.

*Bloat* is a discomfort caused by a build-up of gas, which affects ruminant animals sometimes, if they get on to some unusually rich grazing or something like frosted cabbage. It can be fatal if not treated; in bad cases the vet makes an incision to let out the gas. An incipient case can sometimes be treated with something to lower the surface tension of the foam inside the animal. Margarine can be used, melted and given as a drench (poured down the throat from a bottle; this takes a little skill because a drench must not, of course, go into the lungs.) A recipe from the old books goes as follows :-

*Recipe for Bloat or Hoven*
   1 teaspoon bicarbonate of soda
   1 teaspoon ground ginger
   given in 4 oz warm water.

This sounds quite sensible. The best thing, of course, is prevention of bloat and unless your sheep escape and gorge themselves in forbidden fields, you should be able to avoid it. Always introduce new food gradually, especially when the change is to richer food such as spring grass. Give a feed of hay before moving them onto a fresh patch of grass in the spring.

*Cast Sheep.* Sometimes a ewe will get on to her back, perhaps when rolling to rid herself of the itch caused by parasites, and then by unable to get up again. This is more likely when they are heavy in lamb, fat or unshorn. The ewe will die surprisingly quickly - sometimes in a few hours. A slight hollow in the ground is sometimes enough to get her on her back and keep her there.

A cast or 'rigged' sheep therefore always needs help if you see one, whoever it belongs to. In my travels I have gone to help scores of sheep on their backs, but I have rarely needed actually to touch one. The fact of my approach has often given them that extra push and they have scrambled up on their own. I have sometimes thought that I was wasting my time and that they were not really in difficulties, but it is possible that they would have stayed there until they died.

*Lameness.* This is another thing you will be watching for every day. Sheep grazing on their knees, or hobbling about, are in need of attention. Sometimes it is a case of stones stuck between the claws, but it may be something else. In natural conditions sheep travel a lot on stony ground and their feet are worn down, so mountain sheep usually have good feet. Lowland or back-yard conditions may mean smoother going - and feet in need of paring. The outer shell grows long and may curve round like a nail. It needs to be trimmed back with a knife; get an experienced hand to show you.

*Foot Rot* is a horrible cause of lameness in sheep. It is persistent in the feet and if not treated, it will cause them much pain. Fortunately the bacteria can only live for a few days on the land, so you can get rid of it. The bacterial infection is more likely to get into the hoof in wet conditions, so try to keep the sheep on dry ground. Treatment for foot-rot consists of paring away the diseased tissue (which smells horrible) and treating the hoof with Stockholm tar. A paste used to be made of a quart of warm Stockholm tar, 2 oz copper sulphate and 1 tablespoonful Lysol. This was applied to the foot and covered with a bandage - or a special little Dunlop zip-up boot, which used to be made for the purpose.

When you buy in new sheep or if you are trying to control foot-rot, a formalin foot bath is a good idea. 5% formalin solution (or alternatively 5% copper sulphate bath) helps to harden the feet. Fill a shallow tray with the solution, the tray 6 inches deep and the solution 4 inches deep, arranged so that the sheep must walk through it - say in the doorway of a building. The feet should be trimmed and cleaned up first.

There is now a vaccine against foot-rot, so if you have a problem with it, consult your vet about vaccination.

*Blowfly Strike* is another unpalatable fact of sheep life. In warm weather a certain species of fly, like a bluebottle but smaller and greenish, will lay its eggs on sheep, often in the soiled fleece around the tail. The maggots hatch out and start to eat the sheep; this is a 'strike'. It is why lowland sheep have their tails docked; it lessens the chance of maggots. Mountain breeds are not so likely to be affected and their tails are normally left long.

Shepherds clip the wool round the tails of the sheep in the late spring and remove the soiled wool - doing this in cold weather may give the sheep a chill, so leave it until the weather warms up a little. It is a rather unpleasant job but it makes shearing easier when the time comes, as there will be less dirty wool. But if the sheep are very dirty round the tails, they are scouring and an effort must be made to find out why. If they have been put into a field with a lot of rich grass, over-feeding may be the cause; but worms can cause scouring and they may need worming.

The flies lay their eggs on the sheep, but unless the necessary moist conditions are present, the eggs will not hatch out. If you do get a struck sheep, apply one of the remedies such as Cooper's Fly Dressing after you have cleaned up the area. All maggots and eggs will have to be removed. Keep a watch out for eggs in warm weather and remember that they sometimes occur on the shoulders and back, where birds may have soiled the wool. It is very important to spot this condition and act as soon as it occurs. If you see a sheep restless and wagging its tail, rubbing and biting, look closer. Struck sheep are so unhappy that they rapidly lose condition.

*Shearing.* When the spring weather gets really warm, you begin to sweat for the sheep and to feel that they would be better out of their fleeces. Shearing takes place when the wool 'rises' or ripens. This means that it stands up a little from the skin and the *yolk* - a greasy substance - appears in it. Then shearing is easier; the wool ripens sooner when the weather is warm and when the sheep are well fed, which is why the date for shearing will vary a little from one year to another. This greasy substance in the wool is a mixture of lanolin and suint or sheep sweat.

Shearing is a job which you might like to learn; it would be pleasant to be able to shear your own sheep. But -take instruction! I have sheared one or two sheep, but I would never start to do it on my own. Beginners either cut the skin of the poor sheep, or 'double cut', that is cut too far away from the skin and have to take another swipe. In any case, shearing is very hard work and it stiffens muscles that you never knew existed.

**SHEEP CLIPPING.**
THE MEN TIED THEIR LEGS TOGETHER AND PLACED THEM IN THE LAPS OF THE WOMEN, WHO WERE SEATED ON THE GROUND AND WHO CLIPPED THEM WONDERFULLY WELL WITH HUGE SCISSORS OR CLIPPERS. IT WAS A VERY PICTURESQUE SIGHT, AND QUITE CURIOUS TO SEE THE SPLENDID THICK WOOL PEEL OFF LIKE A REGULAR COAT.
*More leaves from our life in the Highlands.*

You can ask a farmer to have your flock sheared with his, or ask a neighbour in to do yours. The latter may be preferable while the movement restrictions are in force; at the moment, a permit is needed to move sheep anywhere off the premises except for dipping. If a shearer comes in to do it for you, put a nice high barter value on this strenuous work - sheep shearing is no picnic.

If you would like to learn to shear, there are classes as a rule at the county agricultural colleges or their day release centres.

Shearing time varies from May to July, with exceptions in the extreme north and south of Britain. Some sheep are plucked by hand; this happens to the native sheep of the Shetland and Western Islands.

Whether you sell the wool or keep it, the floor where you shear should be very clean and the fleeces should be rolled in the proper way before being stored. Your shearer will show you how to roll the fleece, making a sort of roll with the neck wool to tie it up into a bundle.

If you want to sell your wool and have more than four sheep, you must by law be registered with the British Wool Marketing Board, Clayton, Bradford, West Yorkshire. They will supply you with the inevitable forms. There are over 300 wool grades, including a special one for Jacob sheep.

SHEEP SHEARING.

*Dipping.* This is the next routine job; it rids the sheep of parasites - ticks, keds and lice, any of which can make their lives a misery. Dipping is also compulsory because of Sheep Scab, which we will deal with later.

*Lice* are, in fact, not found if the sheep are properly dipped, preferably as soon after shearing as you can. The wool needs to have grown a little to hold the dip.

The *Tick* is a parasite which sucks the blood of the sheep and falls off onto the grass when full; when it wants another meal it

46

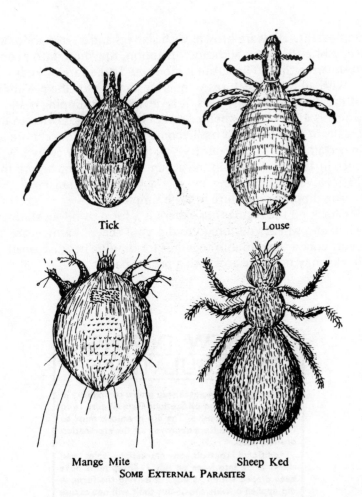

| Tick | Louse |
|------|-------|

| Mange Mite | Sheep Ked |
|------------|-----------|

SOME EXTERNAL PARASITES

catches a passing sheep. Thus it can transmit blood-borne diseases (such as louping-ill) from one animal to another, in addition to causing irritation. Ticks are more numerous in some districts than others, being more often found on rough grazing. This is one reason why hill grazings are burnt off in winter. A really badly-infested pasture needs to be rested for about a year; sheep should be dipped before being moved from an infested pasture to a clean one.

*Keds* are wingless flies, peculiar to sheep and they are very unpleasant. They are brown, with six legs and a large abdomen. They bite the sheep and cause irritation; they hop onto people at shearing time! The autumn dip should take care of keds.

With all these creatures wandering about, plus the possibility of the dreaded blowfly eggs, it is obvious that dipping is a job which is a necessity, whether compulsory or not. Owing to outbreaks of Sheep Scab it has become compulsory once more and in certain parts of the country you have to dip twice by law, once in June and again in the autumn at some time before the 13th November. (See the end of the book for details)

The dipping procedure is quite simple:-

1. Check on the regulations - there is a form to fill in, stating when and where you intend to dip your sheep. Farmers are being encouraged to join together for dipping because small flocks may not have access to a proper bath.

## NEW DIPPING REGULATIONS

Dipping sheep against sheep scab is compulsory in England, Wales and all Scotland from August 16 to November 13 inclusive. The dip chemicals must be Ministry of Agriculture approved for the eradication of sheep scab.

Additional controls on marketing, sales and shows came into force on July 1. Sheep must have been dipped within 56 days of leaving the farm. A dip applied outside the 56-day limit will not excuse the need to dip again during the statutory three months.

Animals for immediate slaughter are excepted, but they must have a local authority licence.

Sheep will be accepted for markets, sales or shows only when accompanied by the owner's declaration that they have been dipped or by a local authority licence. In the latter case, sheep will also have to be licensed for removal from the market.

Contact the divisional vet for further information.

2. Inform the County Council Special Officer who will want to be there when you dip —actually this may be taken care of by filling in the form, but check.

3. Follow the instructions for the dip and use an *Approved Product*. Now this is a tricky subject because by "approved product" most people understand the conventional sheep dips containing Gamma BHC. But there are sheep farmers who are

# WARNING

**Dip is poison... Partly used cans should be carefu... re...aled and stored under lock and ... al... with any full containers, we... ut of r...h of children, animals, and ...vay fro... feed. Empty sheep dip ca... should b... ushed and buried away ... ...livestock. Cover dips so that people and animals do not fall in.**

worried about using this substance (a chlorinated hydrocarbon) which is actually banned in Denmark and Norway. Largely owing to the stand recently taken by one farmer, Mr Ian Neilson, who refused to dip with Gamma BHC, the MAFF have resurrected a pamphlet giving three recipes for sheep dips which may be approved providing the dipping is done twice with exactly eight days between dips. The important thing about these recipes is that they use only natural substances. The MAFF have also supplied a list of manufacturers who might market an alternative type of dip which could be approved. The Soil Association's Quarterly Review of December 1976 contains an article by Mr Neilson and the Soil Association has kindly sent me the MAFF list, which is reproduced at the end of the book. Until now it has not been generally accepted that these other products would be approved, so you may experience local difficulty at first but do stick with it. Start in good time and have a chat with your local Officer about it. I think that this is one small way in which we can help in the battle against pollution.

## PRESCRIPTIONS FOR CERTAIN SHEEP DIPS APPROVED BY THE MINISTER

(Quantities for 100 gallons of dipping bath.)

1. Lime and Sulphur.

Mix 18lbs of sulphur with 9lbs of good quick lime. Slake the lime and make into a thick paste with the sulphur. Place the mixture in a strong cloth, tie the ends and suspend in a boiler containing ten gallons of water so that the water completely covers the contents of the cloth. The cloth must not touch the sides or bottom of the boiler as otherwise the cloth may be burned and its contents escape. Boil for two hours (the boiler being kept covered throughout the period), then remove the cloth, taking care that none of its contents escape into the water, and throw the solids away. Make up to ten gallons again with additional water and put the liquid into a tight drum or barrel. This quantity is sufficient when mixed with water to make a hundred gallons of dipping bath.

2. Carbolic Acid and Soft Soap.

Dissolve 5lbs of good soft soap, with gentle warming, in 3 quarts of liquid carbolic acid (containing not less than 97% of real tar acid). Mix the liquid with enough water to make 100 gallons.

3. Tobacco and Sulphur

Steep 35lbs of finely ground tobacco (offal tobacco) in 21 gallons of water for four days. Strain off the liquid and remove the last portions of the extract by pressing the residual tobacco. Mix the whole extract and to it add 10lbs of sulphur. Stir the mixture well to secure an even admixture and make up the total bulk to 100 gallons with water.

Note that the period of immersion should not be less than one minute in these dips.

## NATURAL DOUBLE-DIPPING TYPE DIPS

| Manufacturer | Product |
|---|---|
| Howard Baker Ltd | Baker's Liquid Sheep Dip |
| Battle, Hayward and Bower Ltd | Battle's Improved Fluid Sheep Dip and Cattle Wash<br>Battle's Fluid Bloom Quality Sheep Dip |
| Cooper, McDougall and Robertson Ltd | Cooper's Utility Sheep Dip<br>McDougall's Liquid Dip (Blue Label) |
| Hilston Manufacturing Co Ltd | Hilston's Perfect Liquid Sheep Dip<br>Perfect Carbolic Dip<br>Moorland (Improved) Sheep Dip Paste |
| Hull Chemical Works Ltd | Kilcrobe 80 Sheep Dip (Double)<br>Kilcrobe 100 Fluid Sheep Dip (Tar Acid) |
| Lestoil Products Co | PEMCO Powder Dip |
| Mirfield Agricultural Chemicals Ltd | Killgerm 31% Disinfectant Dip (Tar Acid)<br>Killgerm Sheep Dip<br>Marsten Sheep Dip |
| Standardised Disinfectant Co Ltd | McLeod's Liquid Sheep Dip<br>SDC Zonda Dip |
| Robert Young and Co Ltd | Springbok Liquid Sheep Dip<br>Blue Label Liquid Dip |

4. Stir at intervals if the solution seems to be settling.

5. Change the solution if the dip gets dirty; if several people are dipping on the same day, this can happen.

6. Give the sheep a rest before and after dipping; it is very bad for sheep to be thrown into a cold bath when they are hot and panting after a journey. And also, overheated sheep can absorb toxic chemicals more quickly —this can lead to poisoning.

7. Lower the sheep into the bath back end first, and make sure each one stays in the dip for at least a minute. Push the head gently under for a short time.

8. Try not to dip when the sheep are wet —it will dilute the solution.

9. Wear protective clothing yourself while dipping; it will save your clothes from getting wet and also protect you from the effects of ordinary sheep dip, which has been found to have a depressing effect on people exposed to it. This is another reason for using one of the alternatives.

10. Remember if you use a "natural" dip that dipping has to be done again in exactly eight days.

SHEEP WASHING,

It should be possible to arrange a do-it-yourself dip in a zinc bath for just one or two sheep, but you would have to persuade the Officer to make a visit to see the performance. I have found that they do not invariably attend. I am assured that it takes two people to dip sheep in a zinc bath; this I can well imagine. But if you have no shepherd neighbours it would perhaps be better to go to a little trouble to arrange this, rather than take the sheep a long distance to be dipped. In this case it would presumably be easier to mix up sulphur and lime or one of the old-fashioned recipes.

SHEEP AFFECTED WITH SCAB.

The Sheep Scab which has caused the dipping regulations to be brought back into force is a mite. It causes irritation to the animals, which rub themselves against fences and thus it spreads to other sheep. Those with scab have ragged fleeces and are bald in patches, but this is not invariably due to scab - it may be a sign of other skin parasites.

*Dogs.* Other people's dogs can be a dreadful problem to sheep owners, which causes me to list dogs among the things which a good shepherd will be looking out for. Respectable household pets can chase sheep through sheer high spirits, meaning no harm; but if the ewes are heavy in lamb, the lambs may be born dead. Sometimes stray dogs will gang up and worry sheep, bring them down and kill them. No wonder some shepherds shoot any strange dog they see near the flock. So this then is another reason why a close watch must be kept on ewes when they are near to lambing, and why it is best to have them right outside your back door at this time if possible. There is also a

53

parasite carried by dogs, a certain tape worm which causes sturdy or gid when it affects sheep. This is not as common as it used to be. The larva gets into a cyst on the brain - which is why sheep's heads must never be fed to dogs.

*Vaccination.* On most farms, routine shepherding includes vaccination against certain diseases. Lamb dysentry and Pulpy Kidney are two of the most common; consult your vet on what prevails in the area.

*Internal Parasites.* These can be a real problem when your sheep are restricted to a small area of ground. As we saw with pigs, land can become 'sick' when one class of animals is kept on it for too long. The old saying ' a sheep's worst enemy is another sheep' refers to this. If you can possibly move them on to fresh ground this is the ideal solution; six months' rest will work wonders for the land. The herbalists use mustard as a cleaning crop, ploughed into the ground as a green manure. Or you can spread the grass with soot and lime to speed up the cleaning of the pasture. If you have a very limited patch, try to winter your animals on a farm; this is quite usual and will give your paddock a rest. The ewes could come back to you in time for lambing.

*Liver Fluke.* This is an internal parasite to beware of in wet seasons. It is a flat worm, reddish and ¾ inch long, which in-fests the liver of the sheep. The eggs pass out into the droppings and in wet fields they get into the shell of the small grey water snail, which is the alternate host. If there are no snails about,

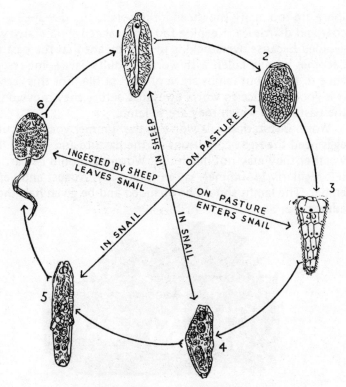

LIFE CYCLE OF LIVER FLUKE
1. Adult Fluke.              4. Sporocyst.
2. Egg of Fluke.            5. Redia.
3. Miracidium.             6. Cercaria.

the eggs die, which is why dry pastures are better for sheep. After development, the young fluke gets onto the grass and is eaten by another sheep. Infested sheep lose condition, develop anaemia and have little appetite. Worming doses now cover fluke as well - both Coopers and ICI have a sheep wormer which deals with all the worms and fluke. Drainage of wet land will also help a great deal. Seasons vary and wet winters are always a dangerous time from the fluke point of view.

*Round Worms.* Some worms are present in all sheep, but when there are too many the sheep loses weight and develops a dull coat and diarrhoea. Keeping too many sheep should always be avoided because overstocking of land means that the pasture becomes heavily laden with worms. Adults have some resistance to them, but lambs are very susceptible. For this reason it is good practice to worm ewes just before they move onto the pasture on which they are to lamb.

Worm infestation gets worse as the summer goes on, but the eggs need the right conditions on the pasture and in very dry weather they may not hatch out. Worming should be carried out regularly in summer, say in June, July, August and September. The lambs should be included and be given half the adult dose.

*Lung Worms.* These are present in the lungs and cause a husky cough, called husk, which you are more likely to hear in the autumn and in damp fields. Infestation can lead to pneumonia. These worms again are more likely to cause trouble when sheep are living close together. On moorland grazings the sheep will be more scattered and so will the worm eggs.

## Clostridial Diseases of Sheep

These are conditions which are brought about by abnormal multiplication of bacteria in the intestines in almost all but not every case. They belong to the Clostridium group of bacteria and can be present in small numbers without doing any harm, like a lot of other pathogens. It seems that the bacteria are more likely to get a hold when a sudden change occurs, such as a change in feeding to better feed.

The vaccines which are given to ewes nowadays are the multiple ones, five or seven in one, to protect against several of the different Clostridial infections. Two or three injections may be given in a year. The lambs are protected for a month or two after birth by the antibodies passed on to them in the colostrum, which is one reason why it is so important that the lamb gets some of that first milk.

*Lamb Dysentery*. This seems to be more common in the northern half of Britain. It kills lambs at about a week old and about half those affected die. They are either found dead, or with severe scouring, often bloodstained.

*Pulpy Kidney*. This is a more widespread disease and affects lambs at an older stage, from six weeks to about four months or even older. It seems to hit the biggest and best and the first indication is a dead lamb.

*Struck*. This is another type of Clostridial infection which suddenly kills adult sheep in very early spring. Two injections are given where it is known to occur, with a fortnight between the two.

*Braxy*. This is induced by indigestion brought on by the sheep eating frosted grass. Ewes can be vaccinated in September against this.

*Mineral Imbalance*. This is the cause of some of the ills which affect sheep. Cobalt deficiency leads to sway-back in lambs. Take care when redressing the balance - these are trace elements and are needed in minute quantities. These problems will be common to the area because they arise from deficiencies in the soil; so the local shepherds are the people to tell you how your soil will affect sheep and whether there are likely to be any mineral problems.

*Twin Lamb Disease.* This is caused by a sort of starvation late in pregnancy and can best be avoided by making sure that the ewes have enough good quality food late in pregnancy.

## Notifiable Diseases

You will hear a great deal about sheep scab while dipping regulations are in force, but this is only one of the notifiable diseases which can affect sheep. These are certain diseases which are particularly dangerous and spread very easily. A suspected outbreak of any one of these in Britain must be reported to the police. If you think one of your animals is suffering from something like this the best plan is to get your vet's advice and he will know the proper steps to take. The list includes these diseases which may affect sheep:

> Anthrax (this may also affect man)
> Foot and Mouth disease
> Sheep scab

In the event of an outbreak, the first task of the authorities is to trace any other animals which may have been in contact with the affected ones. This is why you are required by law to keep an Animal Movement Register. It records the date and identity (ie ear number) of all animals moved onto and off the premises. Every time there is a movement of animals you must enter it in the book; try to keep it up to date, because the police or the local authority Special Officer or the Ministry of Agriculture officials may ask to see it at any time.

Books ready printed for the purpose can be bought, or you can rule out your own in a hard backed notebook.

Many of the problems with sheep are inter-related. Care can avoid the start of the spiral that leads to disease. A sheep that is not quite well is more likely to be afflicted with blow-fly, or a sheep with some kind of mineral deficiency is more susceptible to an attack of worms.

Please do not be put off by this long list of ills. If your sheep are less than fit, a bite of fresh grass will do them good, especially an old herby pasture.

Sheep keeping does tend to be a little hazardous and the vet can be of great help to you. An advantage of sheep is that they can be put into the car and taken to the surgery if need be - just like a pet. This, of course, reduces the fee.

As I said at the beginning, the main thing is to see your flock every day. Anything that does go wrong can then be put right before it assumes disaster proportions.

# PLANTS TO AVOID

| Plant | Usual Habitat and Appearance | Remarks |
|---|---|---|
| Yew ((*Taxus baccata*) | Churchyards, old gardens. A dense evergreen tree. | All parts poisonous. Often kills farm animals in winter. |
| Marsh Marigold (*Caltna palustris*) | Wet boggy land. Yellow flowers in spring and ealy summer. | Not poisonous in hay - toxins lost by drying. Irritating causes blisters. |
| Larkspur (*Delphinium ambiguum*) | Was a garden plant - into fields in many places. Tall blue flowers. | Causes cattle deaths frequently in USA. |
| Wood Anemone (*Anemone nemorosa*) | Woodland in spring. Small white flowers. | Poison: ranunculin. |
| Traveller's Joy (*Clematis vitalba*) | In S. Britain, hedgerow climber, creamy flowers in July/August. | Same as related Anemone. |
| Lesser Celandine (*Ranunculus ficaria*) | Small yellow flowers on banks & hedgerows in early spring. | Ranunculin, very irritating to the skin. |
| Creeping Buttercup (*Ranunculus repens*) | V. common. Like buttercup but with runners. Bad weed in gardens. | All buttercups contain ranunculin, particularly when flowering. Not poisonous in hay. *Aconite* is the really dangerous one of this family. |

| Plant | Description | Notes |
|---|---|---|
| Ivy (*Hedera helix*) | Climbing plant with glossy evergreen leaves. | Berries more poisonous than leaves Sheep can eat leaves in moderation. |
| Hemlock (*Conium maculatum*) | Umbelliferous. V. similar to some harmless plants. Purple blotches on stem, unpleasant smell when bruised. | Very dangerous. |
| Deadly Nightshade (*Atropa belladonna*) | Rather rare. Large purple flowers, June-August. | All parts poisonous. |
| Henbane (*Hyoscyamus niger*) | Grows in seaside places, large yellow and purple flowers June-August. White hairs on stems. Unpleasant smell when bruised. | Introduced by the Romans as a narcotic. |
| Woody Nightshade (*Solanum dulcamara*) | Common hedgerow weed. Purple flowers June-September. | |
| Foxglove (*Digitalis purpurea*) | Tall plant with pinkisk bell-shaped flowers. | Digitalis is a cumulative poison. Not less active in hay. |
| Honeysuckle (*Lonicera periclymenum*) | Climbing shrub with pink and cream scented flowers. | Can produce severe diarrhoea. |
| Lords and Ladies (*Arum maculatum*) | The Arum lily, black rod in a green sheath and scarlet berries | Berries the most poisonous part. |
| Ragwort (*Senecio jacobaea*) | Tall yellow raggy flowers in old pastures. | Still poisonous in hay. |

# 5 growing food for sheep

With plenty of grass at your disposal, you will proably not need to think of growing food for your sheep. Those on a restricted patch may however like to stretch the budget and augment the grazing with some home grown fodder, particularly for the winter months when the grass hardly grows at all.

But before we rip up the land, let us consider grass as a crop, the most important crop there is. To get the most out of grass it must be managed and not just taken for granted.

There are roughly three types of grassland. The first is rough hill grazing, for which the sheep is the ideal customer. The moorland grasses are fine, wiry and low yielding in terms of bulk; they are fescues, bents, nardus and molinia grasses. This kind of grazing suits sheep very well because the ground is usually hard and dry to keep their feet trim; the grasses are short and easy to eat and the sparseness of the grass means that the flock must spread out and wander to get enough food, and this spreads the worm larvae more thinly. Also, of course, the worms are less likely to survive on cold dry ground.

The second type of grassland is permanent pasture; these are old pastures which are never ploughed. Perhaps the land is too steep or too wet or too heavy. This kind of grass can be rather poor, with a lot of weed grasses, or it can be very good

62

indeed. Good permanent pasture will be a dense carpet of bulky grasses and the mixture will contain a lot of perennial ryegrass and some useful herbs.

Permanent pasture is perhaps the best thing to grow if you have only a small plot of land apart from a garden. Never feel guilty about not ploughing up your field and growing crops; under grass, the fertility of your land is safe. The animals grazing it will help to increase that fertility. If the soil is covered with a mantle of vegetation it is safe from erosion and plant food losses, so one could say that permanent pasture is the best crop for a beginner.

The third type of grass is the ley, which is temporary. The field may be sown down to grass for just a year, or for several years. During this time the humus content builds up from animal manure and turf, so that when the field is ploughed there will be a store of plant food ready for sub-sequent crops. This is one way in which organic farmers keep up the 'heart' in their land without using artificial fertilisers.

Sheep can fit in well on all these kinds of grassland. In the main, your management will consist of striking the right balance between undergrazing and overgrazing. Uneaten grass grows long and rough and if you have no more animals to help out the sheep on an undergrazed field the best thing is to 'top' it; this means cutting the grass with scythe or mower, fairly high so that the clumps are cut down. Once they have dried into hay they will usually be eaten, especially if there are nettles among them. An overgrazed field has no grass and should be rested.

A system of small paddocks is the easiest way to manage grass. If there is too much about — and you never know because the yield varies from year to year — then one or two paddocks can be left to grow long and cut for hay or silage. This is better than having to deal with a big, half-eaten field. Paddocks also make resting grass easier. If you have five sheep-grazing paddocks, each one will have a months rest before being grazed again if you use them for a week at a time. This should eliminate all the sheep worms except Nematodirus.

*Silage.* This is grass preserved in acid, pickled as it were, and the acid is the result of fermentation of the grass by bacteria. To get the right amount of heat for the proper sort of fermentation, the heap of silage must be consolidated to keep out the air. This is rather difficult when dealing with a small quantity. A small silage clamp might not be too successful. But there is another method; the grass can be put into polythene sacks, the air excluded and the neck tightly tied. In these airtight containers, the grass will ferment and you will have small parcels of silage ready for use.

When silage is fed, the general rule is that 3lb silage equals 1lb of hay in feeding value but both are variable commodities which makes proper standards difficult. Ewes can eat up to 12lb of silage each per day in bad weather or when there is no grass.

Another rough guide is that 6 to 7lb silage equals 1lb concentrates, but silage should be treated as a bulk feed and used to replace grass. I once worked on a farm where the cows got the best of the silage and the sheep apparently enjoyed the rest, the inferior stuff on the outside of the stack. Our silage was made of lucerne, a useful legume which gives four or five cuts in a season.

Farmers make silage of all kinds of green crops; spare grass is only one of them. It can be made from maize or oats cut green, or a mixture of corn, peas and beans.

*Turnips and Swedes.* These have been traditionally grown for sheep in districts where the sheep were folded on the land for the value of their manure. The roots could be eaten off like this with some good fencing to keep in the sheep, or they could be grown in an odd corner for backyard sheep and carried to them.

Turnips and swedes are probably more popular in areas like the north and west of Britain, with a cool moist climate. They both belong to the Brassica family, so should not be grown for more than a year in the same place, because of the risk of club root; nor should they follow cabbages in a rotation.

Turnips have hairy leaves of a light green, growing straight

out of the bulb. Swedes have a neck from which the leaves grow, and swede leaves are smooth and grey green. Swedes have the higher dry matter content so they are more valuable as a feed. Both can be used as a table vegetable, so you share the crop with the sheep. Some varieties are more suitable for the table than others.

Yield can be about 20 tons per acre and swedes usually have the higher yield. The hardiest swede is Purple Top, popular in Yorkshire and points north. They can be left in the ground until they are needed.

Either of these crops can be sown from mid April until the end of June and in almost any kind of soil except very heavy clay. They like plenty of lime, but not too much, and the land should have plenty of humus in it for good results, especially light land which could dry out in summer.

The seed rate on a field scale is 4lb per acre or perhaps 6lb if the seed is broadcast (scattered), instead of being drilled in rows. On a garden scale, you can sow thinly in drills lft apart and thin out the young plants to about a foot between them. The leaves of the surplus plants can be eaten by stock or as a vegetable.

Some of the less hardy varieties should be lifted before severe frosts and stored in a clamp. Turnips are often lifted in October and swedes in November, but they will go on growing and maturing after this if they are left.

*Kale.* This is a very useful crop and we have fed it to all kinds of animals. There is a good garden variety called curly kale, which makes a small scale crop, or there are larger field varieties. The most suitable field variety for sheep is probably dwarf thousand head, which is low growing and very frost resistant, so it provides food in late winter when the sheep need it most. The yield may be as much as 15 tons to the acre.

Kale provides greenfood when grass is scarce and it helps milk production after the ewes have lambed. It can also be used in the diet of fattening lambs. A suckling ewe will eat about 20lb of kale a day, or if you fed swedes instead, about the same quantity of swedes. To go with this she might need about 1lb of hay and perhaps about 1lb of concentrates as well. Fattening sheep will eat 10 to 12lb of kale a day (commercial farmers expect them to gain 2lb in weight each week when they get over 80lb weight).

*Mangolds.* These are grown in mild climates with plenty of sunshine. They need warmth but otherwise are a remarkably hardy plant, attacked by few pests and diseases. We have grown them successfully in Yorkshire but the yield would have been bigger with a little more sun. They are quite suitable for sheep, but not for us except for wine making.

Mangolds are not resistant to frost, so they have to be lifted in early autumn and stored in a clamp. Cover with straw and then with earth to keep out the frost. They are not ready for feeding until ripe, which is after Christmas, as they mysteriously ripen off in the clamp. My chief

memory of mangold growing is keeping the weeds down in early summer, crawling along the rows on our hands and knees. We sowed them at a rate of about 10lb per acre.in late April and singled them to about 10 inches apart. The yield is variable but you can hope for 20 tons per acre. Mangold tops are not fed to stock.

*Fodder Beet* is a cross between sugar beet and mangolds and has a high dry matter content. The advantage of fodder beet is that it needs no ripening period and can be eaten in the autumn. Also the tops have a good feeding value unlike those of mangolds.

Sugar beet tops, when they have been wilted a week to get rid of the oxalic acid, are good sheep feed if you can get hold of them and it may be possible in beet growing districts.

## Sheep Digestion

The sheep is a ruminant animal; like the cow, it has four stomachs. Ruminants derive much of their energy for keeping warm from the breakdown of fibre in the rumen, and so hardy outdoor sheep will need plenty of hay rather than concentrates in winter to keep them warm.

The natural food of the ruminant is grass, and their stomachs have developed in order to deal with it. In those extra stomachs the fibre in grass (which we cannot digest) is broken down by bacteria into food.

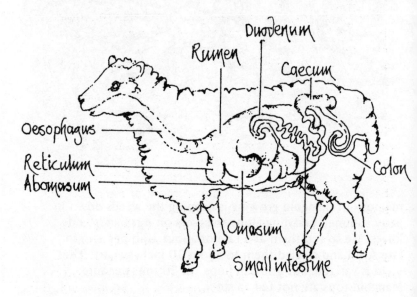

The grass is bitten off close to the ground and is swallowed straight down by the sheep without chewing. It goes first into the rumen, which is the very large first stomach. Then when the animal sits down to rest, it chews the cud. A portion of the grass, called a bolus, is regurgitated back into the mouth and slowly chewed. It is then swallowed and another comes up.

The rumen bacteria work on the fibre much better when

the food is chewed into small particles. When it is fine enough the food is passed on to the other stomachs. The rumen flora, by which is meant the bacteria that live there, develop as the lamb grows; a new born lamb has no rumen action and no bacteria there, and can only digest milk, which is taken by a groove straight to its fourth stomach. The rumen develops as the animal starts to eat solid food and the type of bacteria present will depend on what that food is. For example, an animal on silage will have a different rumen flora to one eating hay. For this reason, any change in the diet of the sheep should be done slowly to give the bacteria time to adjust to the change.

Some food constituents are actually made in the rumen. Vitamin B is produced by the action of the bacteria, as are some amino-acids. This means in effect that ruminants can adjust the quality of the protein they eat and add to the variety of the amino-acids.

From the rumen the food is passed on and the next stomach is the reticulum. This 'reticule' is so called because it collects foreign objects which may be taken in by accident with the food, such as stones or a nail. Sheep are normally dainty eaters but cattle sometimes have to be operated on for the removal of objects from the rumen. The reticulum regulates the water content of the rumen and keeps everything at the right consistency.

The omasum is the next compartment, and the fourth or true stomach is the abomasum. This is where the food, having been broken down by enzymes and bacteria into simple, soluble substances, is absorbed into the bloodstream.

The fibre in young grass is in the form of cellulose which can be digested by the sheep in this way. Older and woody plants have fibre in the form of lignin which is not digestable. This explains why the value of grass varies so much according to the time of the year.

The cellulose is broken down by a process of fermentation, which produces a great deal of methane gas. If this happens to be frothy and the animal cannot get rid of it by belching, as usually happens, 'bloat' is the result.

# 6 breeding sheep

The shepherd's calendar starts in September, the beginning of the breeding year. At that time he starts to think about another crop of lambs.

Ewes should really be given four months' recuperation without lambs before they meet the ram and start all over again. Many people give them rather less; if they lamb fairly late in the spring and the lambs are kept for fourteen weeks on their mothers it can be difficult to fit in four months before it is time to put the ram in again. Also, some people tend to leave the lambs with the ewes for longer than this. But by July, the lambs are nearly knocking over the ewes when they dive in for a drink. They will be eating solid food and will be quite happy by then without milk.

In commercial flocks, they start by inspecting udders and teeth to make sure that every ewe is fit and ready to produce another lamb. This year, there will also be the autumn dip against Sheep Scab to fit in. What conditions should the ewes be in at this time? The answer is - fit but not fat. Get used to handling your sheep and find out how much flesh there is over their ribs. It is not easy to tell by looking at sheep how fat they are - you must handle them.

The idea is that they should go to the ram in 'rising' condition - getting rather better food after being just a little on the thin side, as they naturally will be when they have pro-

70

duced milk all summer. The main thing to remember is that though very fat animals often will not breed, fat ewes have more twins than thin ones.

Some people trim the tail regions of ewes, particularly those with long fleeces, at tupping time to make the ram's job easier.

We have already considered borrowing a ram. He may come with his chest covered in old-fashioned 'raddle' (remember the character in a Thomas Hardy novel who sold this?) so that he marks the ewes as he serves them. Alternatively he may have a harness strapped to him like a parachute, which marks the ewes without spoiling his fleece and also makes it easier to change the colour every three weeks, as is sometimes done. Sheep farmers leave the ram with the ewes for 6-8 weeks, but you can leave him for as long as his owner will allow, because a long drawn-out lambing time will not matter much to you. 21 weeks later you can begin to look for lambs - that is, 21 weeks from his introduction to your ewes.

By now you should have a bunch of pregnant ewes to care for. At first, do not be tempted to feed them too well. Later on in the winter they will perhaps need more food. At the time when the natural food is at its scarcest, in the very early

SHEEP IN WINTER.

71

spring before the grass begins to grow, they may need extra food to help the unborn lamb to develop and also to build up reserves ready to produce milk when the time comes.

Mountain sheep have to forage for themselves. Some shepherds will give them hay in a bad winter, but generally mountain shepherds do not like hand feeding. They say that it discourages sheep from looking for their own food. This is why they do not like too many twins; one lamb is as much as a little ewe on the hill can manage in most winters, and there is more chance of both ewe and lamb surviving if there is only one lamb.

On the other hand, lowland shepherds like two or three lambs and hope for a high lambing percentage. It is important to feed ewes which are expected to produce twins and the food should not be too bulky. Some sort of concentrate will be best. Once again, do not let the ewes get too fat and out of condition. Any animal needs to be fit to give birth easily. Exercise is important, but not rushing around the fields with a dog at their heels! Try to make sure they walk about a fair bit; and make sure that they have a dry bed in damp weather. They can stand cold and wet, but they must lie down some time and if the bed is wet the unborn lamb may get chilled.

When it is time for lambing, you can put them in at night if you have room; this will make the job easier, especially in bad weather.

Once in Wales, I saw a weak light wavering in the corner of a field one dark night; I found a candle in a jam jar stuck in the wall, and round it were ewes about to lamb. This was, my neighbour explained, to frighten away the foxes. Crows can also be a danger during the day; these menacing birds swoop past the lambs and pick out their eyes.

The weather is still the biggest hazard and shelter is important. For a backyard flock it should be possible to rig up pens for them all, to save the difficulty of trying to decide who will lamb first. Sometimes it is easy to tell when a lamb is on its way; the ewe will go off on her own and she will look uneasy. At other times, the baby will be dropped before you know where you are.

Perhaps five in a hundred ewes will need assistance at the

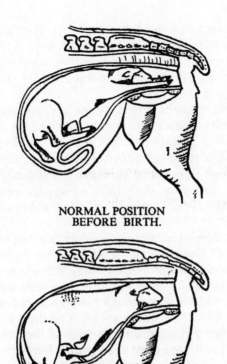

**NORMAL POSITION
BEFORE BIRTH.**

**ABNORMAL POSITION**

time of lambing. Lambs presented the wrong way round may
need turning round before they are born; an experienced shepherd does this job himself, but beginners should get a vet to
see to any lambing difficulties. It does help if you have a
mental picture of how the lamb should look when ready to
come out the right way round. But twins are confusing!

If a ewe has been in labour a long time, really trying and
nothing has been produced, this means that she needs help.
But the general health of the sheep is very important, even
in a difficult lambing. A healthy ewe will recover from a
difficult lambing more quickly and there will be less risk of

infection than when a ewe is not really fit; even so, most vets give an antibiotic injection if they have had to help a ewe to lamb. And after the birth, make sure that she gets rid of the placenta or afterbirth. If this is retained, it may set up infection in the womb.

However many lambing seasons you see, it is always a thrill to get live lambs. They are so frail at first and when things go right, improve so quickly that it is a joy to watch them. They are born wet and must get dry straight away or they may die of cold. The ewe licks her offspring to dry it out with her rough tongue, but when she is lambing outside on a bad night, she can't always cope. Lambing indoors is worthwhile for that reason alone.

And then some ewes, especially when lambing for the first time, will adopt one lamb and leave the other to fend for itself if they have twins. Sometimes they will wander off with the favoured lamb and leave the other in a little wet heap. It is a good idea to have some rough towels handy just in case this happens. The 'little wet heap' must be dried by you very quickly and then put back with the ewe. If the ewe continues to ignore the lamb, you must then get the lamb into the warm. Beside the Aga cooker is the shepherds' favourite.

Some lambs are born with a will to live and they will soon find their way to their lifeline, the teat. Once they have had a drink they are warmed through and have a good chance of survival. Weaker lambs may not bother; they have to be persuaded to drink. Sometimes the ewe will persuade them and sometimes it is up to you. This is a worthwhile job even on a wild night, but of course you have to be there to see what is needed. The satisfaction of saving a little life is a very real one.

Occasionally you get a ewe that never takes to her rejected lamb, be it a single or one of twins, even when you have penned them together for a few days. Sometimes another ewe can be persuaded to adopt such a lamb - either a ewe that has lost her own lamb, or one with a single that seems to have enough milk for two. The age-old practice of skinning the dead lamb and draping it over the adoptive one to give it the right smell worked because the mothers recognised their lambs by smell. However there is the possibility that the dead one died of some infection which the skin could pass on.

In any case of doubt, the ewe should be left in a pen with the lamb or lambs before she is turned out. This is just to check that all is well. You must make sure that the lambs are dry and that they can, and do, drink. Squeeze the ewe's teats to make sure that the milk is flowing freely. A little care at this stage may well save lambs.

When lambing is over, the vigilance goes on. Like all mammals, sheep can get mastitis or inflammation of the udder; this is sometimes caused by a ewe having too much milk for her lamb, or it may be the result of a chill or an injury. Keep an eye on the udders (this is where it pays to have tame sheep, used to handling) and if you find an udder hard or too full, milk it out. But if the milk comes out in clots or the udder is inflamed, there is mastitis present and treatment will be needed. The vet will probably give her antibiotics.

The next job concerns the lambs. It always seems a pity to do it, but they will need to be tailed and the males should be castrated; and the younger they are, the better. Lowland lambs really need short tails because of blowfly attacks and this is really the only way to do it. You can cut off the tail with a knife, but most people now use high tensile rubber rings for detailing. These stop the circulation and eventually the tail drops off.

The same rubber rings (Elastrators) are used for castrating the ram lambs. Get a shepherd to show you how to do it. But - you may be able to manage without castration; either sell the ram lambs quite young, say at about 3 to 4 months, or use them yourself for the freezer, and they need never be castrated. After 5 to 6 months they may need segregating from the ewe lambs, since ram lambs are quite precocious.

Creep feed for lambs is the same idea as for piglets, a food to make them grow faster than they would on milk alone.But in the case of lambs, they do learn to eat grass and some creeps just consist of a hole in the hedge to allow the lambs to go forward onto the fresh grass in the next field before the ewes, since the ewes cannot get through the hole. This 'forward creep' grazing is a very good idea and it also helps to solve the problem of worms, but it is not always possible. There are creep pellets for lambs which are fed in a trough not accessible to the ewes. These feeds are expensive and lambs will grow perfectly well without it. On good grass the lambs will take longer to fatten, but they will be cheaper.

Weaning is at about 14 weeks and the ewes should be taken away from the lambs, so that the young ones are left on familiar ground. Ewes should be out of earshot at this time if at all possible; a lot of pathetic bleating will ensue, but it will not last long.

After weaning the ewes need to be put onto poor grazing for a while, to discourage them from making milk. A watch should be kept for mastitis.

# 7 the harvest

When the lambs are ready, if you have a big freezer why not put in two at a time?

It is best to send them away for slaughter to an abbatoir, but it is perfectly legal to kill them at home if you wish (and it is not so messy as killing a pig), provided that the meat is for personal consumption only and not for resale.

Lambs can be ready for meat from 3 months of age onwards; usually at about 70lb liveweight they would be sold as fat lambs. Mrs. Beeton says that the best flavour comes from a five-year-old sheep, but in those days they liked strong-flavoured mutton. This meat would be tender, because it was hung for quite a while; if you like old-fashioned mutton, kill a mature sheep.

To space out the meat supply, some can be kept on and killed at a more mature age to see what the difference may be. The first lambs could be given creep feed to speed them up - this is usual for early lamb. The later ones could fatten on good grass, with perhaps a few turnips and a handful of oats or barley meal. We fattened two one year on kale, plus swedes and a little 'cow cake'. If you only have a few sheep and do not wish to split the lambs into groups, try the first method one year and the second way the next year, to see which works out the best.

Cutting up the carcase for the freezer is simple enough. You

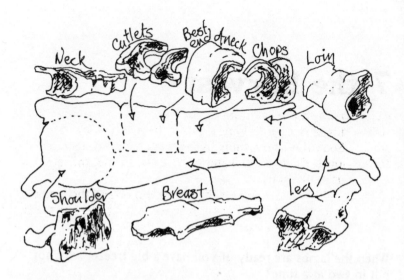

Neck Cutlets Best end of neck Chops Loin

Shoulder Breast Leg

can more or less see where the cuts should be made; and if in
doubt, make it into chops! If you have no freezer one lamb
should present no problem. What you cannot eat will be exchanged for some service (like borrowing the ram that fathered
the lamb) or you could share with a family that kills a lamb
later, and have some of theirs in due course. (This year we intend to swap pork joints for lamb.) Then if there is still a surplus, it can be salted. We will deal with salting later.

Remember to ask for the liver back when you collect the
carcase from the abbatoir; by today's values this is worth quite
a bit. And another thought; why not get back the skin as well?
Usually they will keep the sheepskin, and set it against the
cost of killing and dressing the animals for you. With beef you
get some money back, with sheep it works out about even, so
you get their services for nothing. But it is rather a pity not to
get the skin, and you might like to try making your own sheepskin rugs.

We once found a dead sheep on the road near where we lived
which had been killed by holiday traffic. It was a pedigree ram;
we told the police, and when the fuss had died down and the
farmer had claimed from his insurance company, my brother
was asked to bury the body - so he skinned it first. And then

we tried to cure it. From this experiment we learned quite a lot about how not to do the job; it ended up as hard as a board. A good sheepskin is supple and soft, and the back is like washleather. This is how they should turn out if you do it the proper way.

*Curing a Sheepskin.* As soon as you get the skin back, go to work on it at once and do not leave it hanging about to get smelly. Soak the whole thing in soft water (rain water is good) for 24 hours. Change the water twice in this time; this gets the blood out of the skin.

Next, take the skin out of the water and scrape the inside to get the fat off it. Don't cut it; use something not too sharp,

like a paint scraper.

Boil up one pint of water with one teaspoonful of saltpetre and the same of common salt. Let this cool. Lay your pelt flat with the skin up, and rub it with this salt solution every other day for about ten days.

Then let it dry naturally, in an airy shed. As it dries, keep stretching the skin in your hands from time to time. It is liable to shrink as it dries, and only stretching will make it supple. When you stretch it, the colour changes from a sort of yellow to white.

Next, wash the whole thing in some mild soap. Rinse and

79

let it dry naturally. And then you should have a nice fluffy rug.

*Crooks and Sticks.* While you are in the slaughterhouse you might as well pick up a horn or two — this is the place to find them. a sort of compensation for the unpleasant nature of the visit. In some places horns sell at £4 for half a dozen. If you find a horn you like, you can go in for another shepherd's hobby — making crooks and walking sticks, called stick dressing. Of course your own sheep may give you a horn at some time; but it seems that only ram's horns are really suitable, and the more mature the animal, the better. The horns of ewes have a soft core.

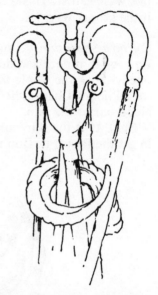

The Dalesman Publishing Company, Clapham, North Yorkshire publish a booklet on how to use horn for sticks and ornaments. 'Shepherds Crooks and Walking Sticks' by David Grant and Edward Hart tells you how to do it. The booklet is an introduction, but if you are interested you may consider competing in the classes at agricultural shows, after learning the craft thoroughly at an evening class; in sheep districts these are now quite common, as interest is growing.

*Macon.* At the beginning of the chapter I suggested that mutton could be salted down instead of frozen. This of course was the main method of preservation in the past, and by the end of winter our ancestors were tired of salted meat.

During the war, this idea was revived and the resulting product was called *Macon* (to rhyme with bacon). Most people heard about it, but nobody was very keen. But recently I talked to a Yorkshire farmer who had made it and liked it.

The chance to make it came when he had to shoot a ram with a broken leg, and decided to preserve the meat for his dogs. So he killed and dressed it, and salted the meat in the same way as for bacon. Then he hung it up without washing off the salt.

After a month or two, he cut off a piece and boiled it for the dogs; it smelt so good he tried it himself. Boiled well, left to cool and eaten cold with salad, it is very like corned beef, he says, but rather nicer. So if you ever have to manage without electricity, this is an idea you might like to try.

*Brine.* Mutton or lamb can also be cured in brine; a recipe I came across recently for curing mutton 'hams' was supposed to be equally good for curing beef, pork, duck and goose! To 1 gallon of water, add

> 1½ lb of coarse salt, preferably sea salt
> 2 oz saltpetre
> 1 lb dark brown sugar
> 2 oz Jamaica allspice
> 1 oz peppercorns
> 1 teaspoon coriander seeds
> 8 crushed Juniper berries

Boil this up for 5 minutes, then cool. Strain and immerse the meat in it. Keep covered in a dry place and under 60 degrees. Leave 10-14 days, depending on size. Wash on removal and soak for 4 hours before cooking.

*Sheep's Head.* Although this dish is now rather out of favour, it is so traditional that we ought to consider it. The abbatoir will give you back the head, but only if you ask for it, so the whole thing can be avoided if you wish! The general method of cooking is to soak overnight in salted water. The salt is washed off and the head boiled slowly for a couple of hours with an onion. When it is cold, the meat can be cut off and minced and you can then proceed to make a civilised dish with it.

*Sheep's Tongues.* These, after soaking in salt water, can be brought to the boil and then casseroled with bacon and vegetables in layers.

*Sheep's Brains.* Mrs Beeton has a recipe for these, once a famous dish. She serves them with a sauce also used for fish, and gives each person one brain.

For Sheep's Brains with Matelote Sauce you will need:

> 6 sheep's brains
> vinegar
> salt
> a few slices of bacon
> 1 small onion
> 2 cloves
> small bunch of parsley
> sufficient stock to cover the brains
> 1 tablespoonful of lemon juice
> matelote sauce

Detach the brains from the heads without breaking them, and put them into a pan of warm water; remove the skin, and let them remain for 2 hours. Have ready a pan of boiling water, add a little vinegar and salt, and put in the brains. When they are quite firm, take them out and put them in very cold water. Place 2 or 3 slices of bacon in a stewpan, put in the brains, the onion stuck with 2 cloves, the parsley, and a good seasoning of pepper and salt; cover with stock or weak broth, and boil them gently for 25 minutes. Have ready some croutons; arrange these in the dish alternately with the brains, and cover with a matelote sauce, to which has been added the above proportion of lemon juice.

For the sauce you will need:

> 30 button onions or shallots
> ¼ lb butter
> ½ teaspoon sugar
> 1 glass of sherry
> 1 dessertspoonful of flour
> 1 gill of gravy
> ½ pint water
> salt, pepper and sugar

Peel the onions and put them, with the sugar, into a 2 pint

saucepan and shake them over the fire, adding the butter gradually. When they are getting brown pour in the sherry and flour and stir gently with a small wooden spoon. When it boils pour in the gravy and water and simmer until the onions are soft, then season with salt and pepper and a little sugar. Strain and serve. The sauce should be of a rich brown colour.

This, then, is the harvest you get when your sheep are killed. Every year we collect from the ewes another harvest; the wool. In the next chapter I will make a few suggestions about the wool, but perhaps in this chapter we can consider sheep's milk, once an important part of the countryman's diet.

*Cheese.* Nobody in this country, as far as I know, keeps sheep now for their milk, because cows and goats will do the job better. Even so, it is possible that one day you could be left with a ewe in full milk and no lamb. While you are looking for a lamb, the sheep will have to be milked, or else the milk supply will dry up. Milking the sheep may well be rather difficult because the teats are so small. But my friend who made the Macon also told me about his sheep's milk cheese; he got the milk out quite well.

— Fabrication des fromages en Suisse. Fac-simile d'une gravure sur bois de la *Cosmographie universelle* de Munster, in-fol., Bâle, 1549.

The Backyard Dairy Book will tell you all about milking and making butter and cheese. With sheep's milk which is rich in fat, it will be important to handle the curd gently. Don't be rough with it or heat it to too high a temperature, or the fat will be lost in the whey.

A quick cream cheese can be made without a lot of equipment and without rennet. Just let the milk go sour naturally in a clean container. When it is solid, pour it onto a square of muslin and gently tie the corners together. Hang the bag over a basin to drain. Next day, take it down and stir in some salt and then hang it up again. When it is dry and crumbly, it is ready to eat. Chives can be mixed in with this curd.

I have made this cheese many times, although not with sheep's milk; it should be delicious. The main thing is to keep the milk really clean, because the 'clean' milk bacteria (Streptococcus lactis and the like) occur naturally in milk and if none of the bugs from dirt get in, which would tend to crowd them out, these good bacteria will complete the process of turning milk into cheese. But they can be discouraged by large numbers of coliform organisms (originating in manure) and you will then get a nasty taste - all sorts of bitter flavours and gas bubbles as well probably. So you have to see that the udder, your hands and the bucket are all clean. If anybody takes the trouble to try this, I would be interested to hear how it turned out.

# 8 wool

Some backyard flocks of sheep are kept just for their wool; other people keep sheep mainly for their meat, with the wool as a pleasant by-product. If you have never handled a fleece before, you may want to experiment with small quantities of wool before you try a whole fleece. You can try out a few of the ideas in this chapter using the scraps of wool that sheep always leave about the field. The fleece could be sold for the first year, say, until you are ready to deal with it. Selling the fleece would be better than storing it for a long time; there is always next year's wool crop.

*Selling Fleece.* You may remember that anybody with more than four sheep, and who wishes to sell the fleece, should be registered with the Wool Marketing Board. After shearing, you take the wool along to the nearest collection centre and it will be graded according to quality and type. The price varies according to the grade, but last year the rough average was £2 for one fleece.

## WOOL FLEECES

The following range of fleeces have been specially selected by the British Wool Marketing Board to cover a wide range of hand-spinning requirements.

| Type | Description | Average Staple Length | Count | Handle | Colour | Approx. Weight of Normal Greasy Fleece |
|------|-------------|-----------------------|-------|--------|--------|----------------------------------------|
| 84227 | Fine Wool White | 3-4" | 56's | Soft | White | 2-3 Kilos |
| 84291 | Fine Wool Dark | 3-4" | 56's | Soft | Grey/Black | 2-3 Kilos |
| 84308 | Romney (Kent) | 4-5" | 50-54's | Medium | White | 3-4 Kilos |
| 84323 | Leicester Cheviot Cross | 6-7" | 50-54's | Soft/Medium | White | 3-4 Kilos |
| 84392 | Jacobs | 5-6" | 50-56's | Soft | Piebald | 2-3 Kilos |
| 84413 | Masham | 4-6" | 46-48's | Medium/Harsh | White | 3 Kilos |
| 94491 | Masham Dark | 4-6" | 46-50's | Medium/Harsh | Dark Grey | 3 Kilos |
| 84603 | Cheviot | 3-5" | 54-56's | Soft | White | 2-3 Kilos |
| 84690 | Welsh | 2-3" | 48-56's | Soft/Medium | Grey/Black | 1-2 Kilos |
| 84719 | Herdwick | 3-6" | Variable | Harsh | Grey/Black | 1-2 Kilos |
| 84631 | Shetland | 1-3" | 54-58's | Very Soft | Browns | 2 Kilos + |

## A GUIDE TO USE

### Staple Length

| | | |
|---|---|---|
| 2" | staple | — for the experienced spinner suitable for woollen spinning |
| 3" – 4" | staple | — recommended for beginners suitable for woollen spinning |
| 5" – 7" | staple | — recommended for beginners suitable for worsted spinning |
| Over 7" | staple | — better suited to the experienced spinner |

### Qualities

| | |
|---|---|
| Soft to medium | — suitable for spinning apparel fabrics |
| Medium to harsh | — suitable for tweeds, coat and upholstery fabrics |
| Harsh | — suitable for upholstery and floorcoverings |

The Wool Board thus has the monopoly for wool buying, but of course if you decide to keep the fleeces, this is your own affair. If you are selling, you will obviously want good grades. A lot depends on fate in the shape of the weather; but

you can help by having them sheared at the right time and by having healthy sheep. Try your best to avoid paint and tar in the fleece. These are difficult to remove and there is a price penalty for them. I used to wonder how the mountain sheep men in Wales fared with wool prices; you could read the owners initials on the sheep half a mile away. There was a system of earmarking with a different notch for each farm, which had been the tradition for a long time and was part of the local history. But it was easier to spot your sheep with a J for Jones on the side; one man went so far as to bathe his ewes from head to foot in rosy pink dye. They were afraid of sheep stealing; rustlers stalked the slopes of Snowdon.

Over forty years ago the Wool Industries Research Association came up with a marking mixture for sheep that was less harmful and would wash out of the fleece. It was rather complicated - a mixture of lanolin, Brazil wax, barytes, a 'suitable pigment' (water soluble dye, I suppose) all mixed up with white spirit.

*Spinning Wool.* Pick up a bit of sheep's wool, pull it out and twist it in your hands. A strong strand is the result. This fact must have been discovered a very long time ago; spinning was one of the first arts learnt by man. Spinning is thus basic, but we are so civilised and specialised that we have grown away from basic things and are just beginning to discover how interesting it is to go back to them; at least, that is how I felt when I started to learn about spinning wool.

Here then is a new avenue ready for exploration when you have wool from your own sheep. It can get quite complicated and need some specialised tools; but at the start, you can spin in the Stone Age way which is still used in primitive parts of the world. Given the wool, you can spin a yarn and knit it up - admittedly into rather a crude garment at first - with no specialised tools other than a spindle, which you can make for yourself.

A simple spindle can be made by sticking a knitting needle through half a potato; this will give you an idea of the size and shape of the tool. You can also make one with a pencil stuck into a ball of clay or plasticene about the size of a golf ball. Our village weaver made me a wooden spindle - you can buy one for just over £1.

First of all, the wool is teased out, which is a sort of fluffing up operation. Take it in your fingers and pull the fibres apart; seeds and bits of twig are taken out and the little knots of wool separated so that a light fluffy mass results. Pull firmly, but not roughly enough to break the fibres. You suddenly seem to have much more wool than you started with.

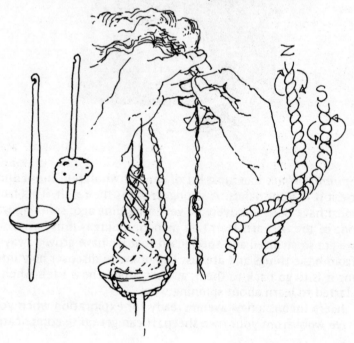

The next step is learning to make yarn. To start off, you wind onto the spindle a yarn already made; thickish wool will do. The weaver who made my spindle showed me how to do it by spinning out a yarn from a lump of fleece in his clever fingers, but if you are starting it will be easier to use woollen yarn. Tie it onto the spindle, loop it round the bottom and up to the top, then loop again. The end of the wool is mixed in with your mass of teased fleece to be spun; draw out a few fibres and wrap them round the yarn.

Spinning is easier if you have seen someone do it. The steps are *Twirl, Pinch, Pull, Let Go.* With your left hand you hold the mass of wool, with the right set the spindle whirling clockwise. The right hand then comes up to pinch out and flatten some of the fibres from the mass between the thumb and forefinger. These are gently pulled downwards; the rotation of the spindle twists these drawn fibres into a yarn. Let go, and go back with the right hand to give your spindle another twirl.

Soon your length of yarn will reach the floor and you will have to stop and wind it onto the spindle shaft. So this sort of spinning is a slow process. It is difficult to imagine doing enough spinning by this method to clothe a family! Yet this is how it always was done, from the time of the ancient Egyptians until the spinning wheel emerged, some think in India. The spindle was the only tool for spinning in Europe until the fourteenth century. When you think of the fine clothes worn by the Greeks and Romans it makes you realise what can be achieved with the spindle.

One advantage of the spindle was that you could walk about while spinning; the women used to spin while walking along behind the sheep and cattle. They must have been so used to the job that they could do it without thinking about it, but, of course, like all crafts it seems difficult at first. The yarn breaks, the spindle clatters to the floor and the yarn is uneven when you first begin. My yarn has loose bits in it where the spindle travelled the wrong way and unwound the yarn. You can spin with either hand and in either direction; clockwise spinning gives you what is called a Z twist, while if you go the other way, you get an S twist.

As you get better at spinning, the length of your draw on

the wool will increase and the twist will be less tight.

SPINNING.

When you wind the yarn round the spindle, wind on more at the bottom than at the top, so that a cone shape of yarn develops. To empty the spindle this is just pushed off. It can then be wound into a skein, with the start and finish tied together in the proper way. Until fairly recently, knitting wool was sold in skeins, which had to be wound into balls of wool.

The foregoing is of course the first step, a way of making yarn on a small scale to find out that you can do it. There are various refinements in the proper art of spinning which may make the job seem more complicated; we will look at a few of the processes in a moment. But perhaps I should mention here that there are several places which offer courses in spinning, and if you want to learn the job properly and make the best use of your fleece, one of these courses would be an excellent idea. In my home area there are spinning courses from time to time at the Adult Education Centre (Grantley Hall, near Ripon). Westdean College, Chichester, Sussex also offer spinning courses and I have also found these two addresses: Craftsman's Mark Yarns, Trefnant, Denbigh, North Wales. The Handweavers Studio and Gallery Ltd., 29 Haroldstone Road, London E.17.

CARDING AND SPINNING, c. 1340.

*Sorting a Fleece.* If you are using a whole fleece, the first job will be to sort it. I mentioned before that the quality varies in different parts of the fleece; if this is sorted into piles of the various types, you will then be sure of using wool of the same kind together, There are all sorts of terms used for the different kinds of wool and they can very in different parts of the country. Sorting wool is a skilled craft; try to get some skilled help to show you how to do it, but failing this, use common sense and a good textbook such as 'Your Handspinning' by Elsie Davenport.

Wool of similar length, waviness, lustre and softness should be put together; use cardboard boxes to sort into. Start with the poorest wool first. The traditional terms for the kinds of wool are as follows:

1. Shoulder (best) — Super Diamond
2. Sides — Extra Diamond
3. Neck (fine and shorter) — Shafty
4. Haunches — Prime Britch
5. Round Tail (coarse, dirty) — Tail or Britch
6. Belly — Picklock
7. Forelegs — Brokes
8. Hind Legs — Britch
9. Back — Diamond

Washing will normally follow unless you are leaving the grease in the wool for hand spinning; wool can be washed at any stage but you cannot dye unwashed wool.

The oiliness of unwashed wool makes it easier to spin, but it must be washed at some stage unless you want to end up with a fisherman's sweater. *Scouring* is the term given to getting the grease out, a rather harsh name for what should be a gentle process. Whether you wash the wool in the fleece, the skein or the 'piece' of finished cloth, do it by steeping in warm water until the water is cold. Then add very mild soap and gently turn the wool over in it, then rinse it in cold water. Do not rub; and never change the temperature abruptly, as you will know if you have ever washed blankets. Wool can stand hot and cold, so long as they are introduced gradually.

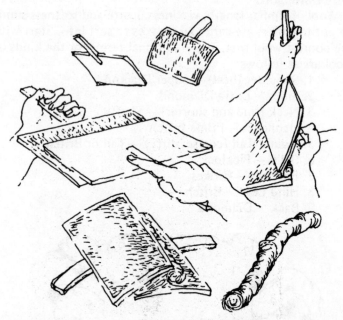

Another process is *carding* or preparing fleece for spinning; just as we teased it out in the hand for a small amount. Carding is really arranging the fibres so that they lie parallel, like combing your hair. This process makes it easier to draw out regularly and so spin an even yarn. The name comes from the French word for teasel, since teasels were used for this job. Carders are two wooden bats with handles, like table tennis bats, which have fine wire hooks on one side. They are marked Left and Right when new, and are always used in the same hand. The stroking action of one carder over the other combs out the fleece and then rolls it up into *rolags* ready for spinning.

You could make a pair of carders yourself. The bats should be of ash or thick plywood, 6 inches wide and 4¼ inches tall. Scratch Card is stuck onto one side with glue and staples round the outside; this card can be bought in some Do-it-yourself shops. It is material with small metal thongs embedded in it.

*The Spinning Wheel.* There are two kinds of spinning on the wheel, woollen and worsted; the latter is the method used for long and silky wools. The cloth thus produced is smoother than

THE SPINNING-WHEEL.

woollen cloth.

Spinning with a wheel is rather involved for a written explanation, and a long list of the parts of the wheel would be boring without a wheel in front of you. The best thing is to learn spinning on the wheel from an experienced spinner, such as you will find on the courses I suggested earlier; and learn about it before you think of buying a wheel. In the matter of dealing in spinning wheels, there are pitfalls for the unwary. A wheel is bound to be expensive and you need to know what you are buying.

Saxony Spinning-wheel.

Eliza Leadbeater is a person who could help if you want to take up spinning seriously. She has a shop and a world-wide postal service for what seems to be all the requirements for home spinning and weaving. Her address is Rookery Cottage, Dalefords Lane, Whitegate, Northwich, Cheshire. Her catalogue is like a house magazine, with a nice personal touch, and I have found her service prompt. She also runs classes in spinning.

Another source of spinning wheels is Malcolm MacDougall, Grewelthorpe Hand Weavers, Grewelthorpe, Ripon, North Yorkshire. His prices range from £35 to over £100 for a new wheel; he has a small number of beautiful individually made wheels because he knows a man who makes them for a hobby.

Further south, spinning wheels and other equipment for home spinning and weaving are available from Frank Herring and Sons, 27 High West Street, Dorchester, Dorset.

# 9 using your wool

Your spun wool is now ready for use; at first it may be thick and rather uneven but even like this it has, literally, a home-spun charm. It can now be knitted or woven into something either useful or decorative, or both.

*Knitting.* There is now a theory put forward by some research-ers that knitting is even older than weaving, and that the first clothes which were made from spun wool were knitted up on bone or wooden needles. In certain rural museums there are long, pin-like objects found on prehistoric sites, which have been labelled as tools for cleaning out the ears. Now, they are suspected to be knitting needles.

It is possible that the patterns early knitters used had some kind of magic value to the garment's wearer. Certain combin-ations of numbers were thought to keep away the evil eye and the result of putting them in the pattern and knitting 7 plain, 3 purl was a thick ribbed effect. This pattern was apparently common until the 12th century.

Knitting's image has degenerated somewhat to its present cosy fireside female job whereas at one time it was an occup-ation for men, like weaving.

The only problem when you knit up your homespun wool will be to use a pattern intended for the same thickness of wool. Your sample will need to be compared with the different sorts of knitting wool so that you can decide how it should be knitted up. One strand from the spindle is one ply, two twisted together will be two ply; is you want to produce a yarn like bought three ply, your yarn will need to be quite fine and then three strands of it will be twisted together on the spindle or the spinning wheel. It will soon become apparent, if your first attempt is a scarf or a woolly hat, how the spinning should be modified to produce knitting wool. With practice your yarn will get finer, but the fineness will depend to some extent on the type of fleece you are using.

*Weaving.* This is another fascinating art and if you use a whole fleece you may want to try weaving your yarn. The principle is simple. Whereas knitting involves looping one continuous thread, weaving is the interlacing of two sets of threads at

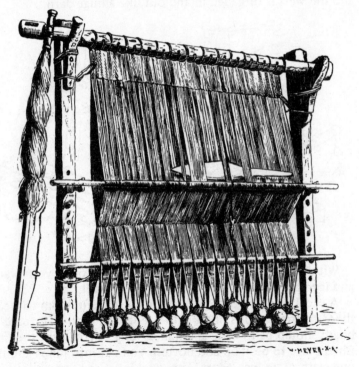

LOOM FROM FÆROE ISLES.

right angles to each other. The vertical set is called the *warp;* these must be tough because they tend to take much of the strain of the cloth. The other set of threads is the *weft.*

The loom is the frame which holds the warp tight so that the weft can be threaded through it. Looms can be huge and expensive or they can be simple and reasonably cheap to make. Many things can be used to make a frame; I will mention some of them and if you want to know more, consult one of the books on weaving which are now coming on to the market as a result of the growth of interest in the craft.

Branch weaving is one of the more primitive ways in which to make things like cushion covers. It was used by the North American Indians. You get a rigid branch and a pliable one, and bend the latter round in a U shape to meet the rigid branch at both ends. Within this frame the vertical threads are strung and the weft is threaded in and out like a huge darn.

With a backstrap loom, you pass one end round your waist and lean back against it to keep the work taut.

A sheet of stiff cardboard will make a useful loom to weave the yarn from your spindle into a length of cloth. The cardboard should be as wide and half as long as you want the finished piece to be. Top and bottom you cut into jimps about one-eigth of an inch apart. The wool is wound round the card-

WARPING FRAME.—(*Woollen Yarn.*)

board and held by the jimps to give the warp, and the weft is darned in.

The principle of the proper loom is that it makes the under and over movement of the weft automatic. It does this by picking up every other warp thread and leaving the rest, thus creating a space between them which is called the shed. The weft is passed through this to save the laborious darning process.

The plain weave, like darning, is called tabby weave. Basket weave is made by raising the threads in twos. There are many other different weaves, making the textural differences of cloth. The darning needle is of course replaced by the shuttle.

At school with our small table looms, we found it difficult to keep the weft tight, and we ended up with rather straggly lines of weaving wandering up and down, and a strangely shaped piece coming off at the finish. I think it would be because we did not push the filling up tight after each row, as we should have done. This is done with a 'beater' - a comb or a table fork will serve the purpose. See that the line is even before you do another one.

Weaving proper is a very mathematical process. A weaver told me that even his most random patterns were in fact worked out in detail before he started, whether the pattern was in the weave itself or in the colours he was using.

101

There are people who have started by weaving wool, and have progressed to divesting their pets of surplus hair ( at least I hope it is surplus!). There are hairy dogs that produce enough fibre to be spun and woven; cats and Angora rabbits. Archaeologists used to think that the prehistoric weavers mixed deer hair with their wool, but they now suppose that the few shreds of rather harsh material which have come to light have been made just from the kempy wool of the little prehistoric sheep.

One of the books you ought to look at is Irene Waller's 'The Craft of Weaving' published in the Craftman's Art series by Stanley Paul. This book takes the beginner step by step through the process of simple weaving. It tells you exactly how to make a weaving frame, which the author describes as a middle way between the primitive weaving and expensive equipment. This book is both practical and artistic and it may give you ideas. It suggests that you can weave with unspun fleece.

Common Hand-loom.

*Dyeing.* Once you have done some weaving with plain wool, it may occur to you how pretty it would look with a weft of a different colour, or if the work had bands of colour running through it. If you have Jacob sheep, they will provide you with some variety of colour; but for white wool, why not try dyeing.

Dyes are of two main types, natural and synthetic. If you want to be sure of clear, bright, fast colours, the synthetic dyes are best. You can buy most of them from ICI Dyestuffs Division, P.O.Box 42, Hexagon House, Blackley, Manchester. Or you could try a firm which specialises in selling small quantities, of which nobody could accuse ICI! They are UNI-DYE, Rear Castle Yard, Church Street, Ilkley, Yorkshire. Within the synthetic range there are acid, direct and reactive dyes.

Natural dyeing is not so predictable and is more of an adventure. However, it is a prime backyard pursuit. Many common plants can be used to give colour to your wool and it is exciting to experiment with small quantities until you find the shade you are looking for.

Weld: usea for dyeing yellow.

Most vegetable dyes will wash out again, as you know if you spill beet-root on the tablecloth. But you want the wool dye to be safe or washable, and not to fade in the sun. So the dye is fixed with a mordant in some cases, which is a mineral ground for the dye to stick to. Some plant materials for example walnut bark and lichens — do not need a mordant and they are called substantive dyes. The others, which need a fixative, and are called adjective, are more common. The same dye will sometimes give different colours with different mordants.

Most of the minerals used for fix-atives can be easily obtained from the chemist, but Eliza Leadbetter whom I mentioned earlier, supplies dyeing mat-erials, so if you have any difficulty, ask her. Salt, vinegar and rusty nails were favourite mordants with traditional home dyers, although the process must have been a hit or miss one often. Salt often makes the colour brighter.

Woad: used for dyeing blue

You do not always get the colour you expect; red dahlias, for example, give a yellow dye. Apple and pear twigs also dye yellow.

Dyeing can be done in the fleece, the skein or the piece of woven material, but for spinning it may be easiest to dye the skein, so that you can spin the unwashed wool. On the other hand, little bits of fleece will be best for early experiments, and if it is dyed in the fleece you can spin together strands of differently coloured wool.

First wash your wool, at whatever stage it is, so that the dye will stick. Use, as I said before, gentle soap and soft water.

*Mordanting.* This is done before the dyeing process, so that

Madder · used for dyeing red.

there is something for the dye to stick to. The mordant must
be properly dissolved in the water before the wool is added.

For three gallons of soft water, rain water if possible, the
following quantities will be about right:

    1½ cups salt
    OR 3 oz copper sulphate
    OR 6 cups vinegar
    OR 3 oz alum

The water is lukewarm; after the mordant is dissolved, the
fleece or wool is put into it. Then bring the whole lot to the
boil and immediately turn down to a simmer. Simmer for an
hour, then fish out the wool and rinse in HOT water. Do not
use more of the mordant than necessary because chemicals
are obviously not good for the wool. After this the fleece is
drained, and as you dye fleece when wet, it can be popped
straight into the dyebath.

The use of iron as a mordant is rather different ( the modern equivalent of rusty nails) because it is used in the middle
of dyeing instead of at the beginning. To 1 lb of wool use ½

oz of iron, 2 oz of cream of tartar and 4 gallons of water. The wool is put in the dyebath and boiled for half an hour. The wool is then taken out and the iron mordant mixture added to the dye. Then the wool is put back and simmered for another half an hour, which is known as 'saddening'.

*Dyeing.* To prepare the dye, the plants are mashed up and soaked in water at least overnight first. Tough things may need several days or even weeks. There are several ways of doing it. Roughly 1 lb of material will be used for every 1 lb of wool. One way is to crush up the plant material, and put it in a stainless steel container; pour boiling water over it and leave for three days. Stir it up each day. Then strain the water off into the dyebath, and simmer the wool in it for an hour. If there is not enough dye solution to cover the wool, add clean water to top it up. Another way is to steep the plants in household ammonia for three days - the colours will be brighter, but the smell will be vile.

Brazil-wood: used for dyeing red.

It is not essential to take out the plants from the dye, they can be left in and boiled up with the wool. One pound of wet wool per three gallons of water is right for dyeing, so to do it in bulk you may need a bucket on top of the stove, or a big cauldron.

Only natural fibres will take the natural dye, so if you twist in a synthetic fibre with your wool this will not be harmed but will emerge undyed. Colours will vary a lot, so it may be hard to match one batch with another. Therefore try to do all you will need at one go. There are variations according to the time of year and the place in which the plant is picked. A record of what you do will help you build up experience.

Turmeric   used for dyeing yellow

# PLANT DYEING TABLE

The colours given are a general indication because so many shades and variations are possible. The mordants listed are suggestions only; some plants don't necessarily need one e.g. oak, which is high in tannin, will give brown on its own, but with alum gives gold. Most lichens and some plants, notably woad, will not easily give up their dyes unless fermented and/or boiled. The traditional method of fermentation was to leave the dye plant for several weeks or months (in some cases years) in urine. Modern equivalents are ammonia or lime. The key words here are 'Experiment for yourself'. Much of the traditional knowledge of dyeing has already been lost, and is waiting to be rediscovered.

The common names given are those common in the U.K. In the U.S.A. there may be different so please refer to the Latin names if in doubt. ("Two nations divided by a common language" — G.B. Shaw.)

| Common Name | Latin Name | Parts Used | General Colour Guide | Suggested Mordants |
|---|---|---|---|---|
| Agrimony | Agrimonia eupatoria | leaves | gold | alum, chrome |
| Alder | Alnus spp | bark | yellow/ brown/ black | alum, iron copper sulphate |
| Alkanet | Anchusa tinctoria | roots | grey | alum, cream of tartar |
| Apple | Malus spp | bark | yellow | alum |
| Barberry | Berberis spp | twigs | yellow | alum |
| Bilberry | Vaccinium spp | berries | purple | alum, tin |
| Blackberry | Rubus spp | berries, young shoots | pink, purple | alum, tin |
| Blackcurrant | Ribes spp | berries | grey, deep purple | alum, tin |
| Blackwillow | Salix nigra | bark | red, brown | iron, chrome |
| Bloodroot | Sanguinaria canandensis | roots | red | alum, tin |
| Bracken | Pteridum aquilinum | young shoots, old tops | yellow, green | alum, chrome |
| Broom | Cytisus spp | flowering tops | orange yellow | chrome, tin |
| Buckthorn | Rhamnus cathartica | twigs, berries | yellow brown | alum, cream of tartar, chrome tin, iron |

| Common Name | Latin Name | Parts Used | General Colour Guide | Suggested Mordants |
|---|---|---|---|---|
| Cherry | Prunus spp | bark | pink, yellow, brown | alum |
| Coreopsis | Coreopsis tinctoria | flower heads | yellow, orange | chrome, tin |
| Cypress | Cypress spp | cones | tan | alum, chrome |
| Dahlia | Dahlia spp | petals | yellow, bronze | alum |
| Day Lily | Hemerocallis spp | flowers | yellow | alum, tin copper sulphate |
| Dog's Mercury | Mercurialis perennis | whole plant | yellow | alum |
| Dyer's Broom | Genista tinctoria | flowering tops | yellow | alum, chrome |
| Elder | Sambucus nigra | leaves, berries, bark | yellow, grey | iron, alum |
| Golden Rod | Solidago spp | flowering tops | gold | alum, chrome iron |
| Heather | Erica spp | tips | yellow | alum |
| Horsetail | Equisetum spp | stalks | green | alum, copper sulphate |
| Hypogymnia lichen | Hypogymnia psychodes | whole lichen | gold, brown | |
| Ivy | Hedera helix | berries | yellow, green | alum, iron |
| Lady's Bedstraw | Gallium boreale | roots, tops | yellow red | alum, chrome iron |
| Larch | Larix spp | needles | brown | |
| Lily of the Valley | Convallaria majalis | leaves | gold | lime |
| Lombardy Poplar | Populus nigra italica | leaves | yellow, gold | alum, chrome |
| Madder | Rubia tinctoria | whole plant | orange red | alum, tin |
| Maple | Acer spp | bark | tan | chrome, copper sulphate |
| Mahonia | Mahonia aquifolium | roots, berries, whole plant | blue, brown | alum, chrome |
| Marigold (Try also Tagetes spp) | Calendula spp | whole plant flower heads | yellow | alum, chrome |
| Meadowsweet | Filipendula ulmaria | roots | yellow, green | alum, iron |

| Common Name | Latin Name | Parts Used | General Colour Guide | Suggested Mordants |
|---|---|---|---|---|
| Menegussia lichen | Menegussia pertussa | whole lichen | pink, when boiled | washing soda |
| Nettle | Urtica dioica | fresh tops | yellow, green, grey | alum, iron |
| Oak | Quercus spp | inner bark | gold, brown | alum, chrome |
| Ochrolechina lichen | Ochrolechia parella | whole lichen | orange, red when fermented in urine, then boiled | alum |
| | Ochrelechia tartarea | whole lichen | red, purple when fermented in urine, then boiled. | alum |
| Onion | Allium cepa | skins | yellow, orange | alum |
| Parmelia lichen | Parmelia caperata | whole lichen | yellow, brown, when boiled | oak bark |
| Peltigera lichen | Peltigera canina | whole lichen | yellow when boiled | alum |
| Pokeweed | Phytolacca americana | berries | red, tan | alum |
| Privet | Ligustrum vulgare | leaves, berries | yellow, green, red, purple | alum, chrome tin |
| Pyracantha | Pyracantha angustifolium | bark | pink, brown, grey | alum, chrome |
| Ragwort | Senecio | flowers | deep yellow | |
| Silver Birch | Betula pendula | leaves, bark | yellow, gold | alum |
| Sloe (Blackthorn) | Prunus spinosa | sloe berries, bark | red, pink brown | alum |
| Snowberry | Symphori-carpus albus | berries | yellow | alum |
| St. John's Wort | Hypericum spp | flower tops | red, yellow | alum, chrome |
| Sumach | Rhus spp | berry tops leaves | tan, brown | alum, chrome |
| Sweet Woodruff | Asperula odorata | whole plants | red, pink | cream of tartar |
| Tansy | Tanacetum vulgare | flowering heads | yellow | alum |
| Usnea lichen | Usnea barbata | whole lichen | yellow when boiled | |

| Common Name | Latin Name | Parts Used | General Colour Guide | Suggested Mordants |
|---|---|---|---|---|
| | Usnea lirta | whole lichen | purple when fermented in urine | |
| Weld (wild Mignonette) | Reseda luteula | whole plant | olive green | alum, cream of tartar |
| Woad | Isatis tinctoria | whole plant | blue | lime |
| Yellow Flag Iris | Iris pseudaconus | root | grey, black | chrome, tin, iron |
| Xanthetia | Xanthetis parietina | whole lichen | purple blue | |

We are grateful to Katie Thear, Editor of Practical Self-Sufficiency, for permission to reproduce the above Plant Dyeing Table. This excellent magazine is published by Broad Leys Publishing Company, Widdington, Saffron Walden, Essex CB11 3SP and costs £3.50 for an annual subscription (six copies).

# 10 one or two ideas

To finish up with, I would like to mention in a little more detail one or two ideas that cropped up in earlier chapters.

From time to time I have said that when choosing a breed, you might like to consider one of the rarer breeds which are fighting for survival. There is now a society devoted to looking after the interests of our rare breeds of livestock. In my opinion this is a very good thing, although there is some opposition to the project by people who point out that the dinosaur is now extinct - so what?

It has worried me for a few years that cattle, for example, are being concentrated by the widespread use of A.I. into very few blood lines or families and fewer breeds. Breeds that were quite common a few years ago are becoming rare almost without our realising it. We used to keep Saddleback pigs, but I have not seen one for a few years now.

The Society which is coming to the rescue is called The Rare Breeds Survival Trust, c/o The Ark, Winkleigh, Devon. It is frightening to be told in the leaflet put out by them that since the Second World War, over twenty breeds have become extinct. The Lincolnshire Curly Coat pig and the Norfolk Horn sheep are examples. It is too late to save these breeds, but the Trust hopes to prevent a headlong gallop away from the minority breeds so that extinction does not happen to so many more.

Jacob sheep are an example of a breed which had reached a dangerously low level, but which is now becoming popular, quite well established and on the increase. So the Trust maintains a 'watching brief' on a breed in this position and hopes that help will not be necessary. And now there is a Jacob Sheep Society they should be safe.

Jacob's ewes are only one of over 60 breeds of animals the Trust is keeping an eye on at the moment. The members are worried about the loss of genetic resources when a breed disappears (plant breeders feel the same about weeds). So the Trust have established an A.I. semen bank in which to store genetic material to help breeders; there may be qualities of hardiness or temperament needed in the future which only fresh blood will provide - and here it will be, waiting.

An interesting project carried out by the Trust was their rescue operation on the North Ronaldsay sheep. This breed, natives of the island of that name, are kept from the precious grazing by a wall all round the island and live on the seaweed they pick up from the shore. But they were in danger and it was possible that this would be the next breed to disappear. The dangers that overshadowed them were the possibility of foot and mouth disease, and also the contamination of the seaweed from oil slicks that come ashore.

The Trust bought an island for a number of North Ronaldsay sheep; Linga Holm, not very far from their original home. Some stayed on North Ronaldsay, but the establishment of a second colony spread the risk of the extermination considerably. It is hoped that on Linga Holm they will be able to breed in safety.

There is now a registration programme. Breeds which have no Breed Society of their own with which to establish a pedigree will be recorded as a way of keeping track of blood lines and it will also keep account of numbers. Already published are two volumes of a Combined Flock Book which covers ten minority breeds of sheep.

There is even an annual Rare Breeds Show and sale, which is held at the Royal Showground, Kenilworth, Warwickshire. What a day for enthusiasts that will be!

If you want to join the Rare Breeds Survival Trust it costs

a minimum of £3 per year. No doubt they will help anybody by putting them in touch with breeders of whatever sort of stock they may be looking for.

The idea is to preserve old breeds for use and for the contribution they can make, not just for exhibition or even solely as a genetic bank. Farmers are being encouraged to keep these cattle, sheep and pigs on commercial lines and record their performance. As a backyarder, you might do well to try, say, Gloucester Old Spot pigs, or Tamworths. Or you might get one of the rarer breeds of sheep. It would certainly make life more interesting!

While I am on the subject of conservation, you may be interested in another organisation which has recently been formed to try to reverse a regrettable trend. Large scale farming has changed the countryside so much that damage has been done. Hedges have been taken out, ponds drained and coppices cut down in the interests of agribusiness. Poisonous sprays have proliferated.

For years, conservationists have wailed about things like this, but mostly to each other. Not until they began to get involved with the people on the land did things begin to happen. The recently-formed Farming and Wildlife Advisory Group contains farmers as well as conservationists; in fact it is run by an ex-farmer. The message is getting over to farmers and since you will probably be a land user, I thought that you might be interested in what the Group suggests.

They are interested in the art of the possible; they want farming and wild life to live side by side. The main theme of their message is that every land user can help in a small way to provide cover for animals and birds in a corner that is too small or uneconomic to be farmed to the hilt. With plenty of cover, wild things can look after themselves. In a farming magazine a few months ago there were reports about delighted farmers who had tried to provide shelter and were thrilled to see birds they hadn't seen for years, such as owls, coming back to live on their land.

So - if you have a bit of land, think twice before you cut down a thicket of bushes in a corner of a field. You will soon be sharing your patch with all sorts of interesting creatures.

We have been conditioned into tidiness and being ashamed of weeds, but it is relaxing sometimes to let nature take its own course on a small area and see what happens. Hedges are very important; they are the corridors down which the wild creatures move from one patch of cover to the next.

With pigs and sheep on a small patch, keeping a balance with the environment may take some effort. Overstocking is to be avoided at all costs; think of the worms. And it is also bad for the land. I have seen upland grazing eaten off to the bare soil by too many sheep. When heavy mountain rain fell, there was nothing to hold the soil and it was washed away, a little more every winter. I am sure that the craggy hills of Western Britain are largely man-made deserts; on thin soils the balance is very delicate, and too many animals can destroy it.

# reference

## General

*The Agricultural Notebook.* Primrose McConnell. Iliffe Books.
First published in 1883, this has been rewritten and brought up
to date. It is a very handy little reference book of facts and
figures for livestock and crops.

*Rations for Livestock.* MAFF Bulletin. HMSO.

*Self-Sufficiency.* John Seymour. Faber and Faber 1973.
All about backyarding; a small section on pigs and a mention
of sheep. Very readable, as are all his books.

*Cottage Economy.* William Cobbett (first published 1823).
He urges all cottagers to keep a pig. Reprinted Landsman's
Bookshop 1975 - a fascinating book.

*Modern Farming.* Waverley 1950. There are several volumes of
this useful work but Vol 2 deals with pigs, sheep and health.
This is agricultural science just before it was taken over by big
business.

*Farming and Wildlife.* Ed. Derek Barber. RSPB 1970.
Conservation and what we can all do to help.

*Factory Farming.* Ed. Bellerby. BAAS.

*The Complete Book of Self Sufficiency.* John Seymour.
Faber and Faber 1976

*Fat of the Land.* John Seymour. Faber and Faber. His best
book and the earliest.

*Brave New Victuals.* Elspeth Huxley. Chatto and Windus.

*The Farming Ladder.* G.Henderson. Faber and Faber. Good
sound farming principles.

*Owner-built Homestead.* Ken Kern. Charles Scribner and Son,
USA.

*Soil Association booklets.*
   Farming Organically
   Smallholder Harvest
   Self Sufficient Smallholding
   Use Your Weeds

*Observer's Book of Farm Animals.* L.Alderson. Warne.

## Buildings

*Build Your Own Farm Buildings.* Frank Henderson.
*Owner-built Home.* Ken Kern. Charles Scribner and Son, USA.

*Owner-built Homestead.* Ken Kern. Charles Scribner and Son,
USA.

*Shelter Belts and Windbreaks.* J.Caborn.

## Sheep

*Profitable Sheep Farming.* M.McG Cooper and R.J.Thomas.
Farming Press.

*Sheep Farming Today.* J.F.H.Thomas. Faber 1966. This book is interesting and readable.

*The Sheep.* Youatt. A nineteenth century book, a classic in its day and much quoted by the old literature. You may pick one up second-hand. I am still looking for mine!

*Shepherds' Crooks and Walking Sticks.* Grant and Hart. Dalesman Books 1975.

## Health

*Herbal Handbook for Farm and Stable.* J. de B. Levy. Faber and Faber 1975. A lot about herbal treatments for sheep and full of useful advice. A very good book for backyarders.

*TV Vet Sheep Book.* Farming Press.

*Diseases of Sheep.* V.G.Cole. Angus and Robertson 1966.

## Crafts

*Textile History.* David and Charles. A volume is published each year; very good background stuff.

*Country Bazaar.* Pittaway and Scofield. Fontana/Collins 1976. This is full of ideas and addresses, a good source of information about many country crafts and pursuits.

*Your Handspinning.* Elsie G. Davenport. Select Books 1964. Very clear instructions.

*Handspinning.* Eliza Leadbetter 1976.
A new book on the subject.

*The Craft of Weaving.* Irene Waller. Stanley Paul 1976.
A good book for the beginner.

*Natural Dyes and Home Dyeing.* Adroska. Recipes and historical notes.

*Use of Vegetable Dyes.* Violetta Thurston. Dryad Press.

*Vegetable Dyes.* E. Mairet. Faber.

*Card Loom Weaving.* Dryad leaflet.

*Handweavers Source Book.* Davison. A reference book.

*Simple Weaving.* Hilary Chetwynd. Studio Vista.

*Weavers Book.* Tidball.

*The Weavers Craft.* Simpson. Dryad Press.

*Cloudburst 2.* Ed. Vic Marks. Cloudburst Press, Canada. Designs for building your own carding machine, spinning wheels and looms. An excellent book, but difficult to obtain. May be available from Compendium Books, London.

## USEFUL ADDRESSES

### Suppliers of equipment

Self Sufficiency and Smallholding Supplies,
The Old Palace,
Priory Road,
Wells,
Somerset.
A converted cinema which contains everything that the backyarder might need. Send £0.50p for their comprehensive illustrated catalogue.

### For craft equipment

Eliza Leadbeater,
Rookery Cottage,
Dalesfords Lane,
Whitegates,
Northwich,
Cheshire.
All spinning and weaving equipment in stock.

Malcolm MacDougall,
Grewelthorpe,
Ripon,
North Yorkshire.
Spinning and Weaving equipment.

Frank Herring and Sons,
27 High West Street,
Dorchester,
Dorset.
All spinning and weaving equipment together with many
other craft supplies.

Handweavers Country-Style,
Northfield,
Vermont 05663,
USA

The Makings,
2001 University Avenue,
Berkeley,
California 94704,
USA.

## Education

Westdean College,
Chichester,
Sussex.
All kinds of skills taught including spinning.

Soil Association,
Walnut Tree Manor,
Haughley,
Stowmarket,
Suffolk.
Short course on organic husbandry.

Working Weekends on Organic Farms (WWOOF)
A good idea for beginners to find out what they would like
to do. Contact the Soil Association for details.

For short privately run courses see the small ads section in
Practical Self Sufficiency,
Broad Leys Publishing Company,
Widdington,
Saffron Waldon,
Suffolk.


# ORGANISATIONS OF INTEREST

**General**

The Soil Association,
Walnut Tree Manor,
Haughley,
Stowmarket,
Suffolk.
Advice given on organic gardening and farming.

The Henry Doubleday Research Association,
Convent Lane,
Bocking,
Braintree,
Essex.
As for the Soil Association.

Both of these are small organisations but with international
membership. They give a great deal of help to people like us
and are well woth joining.

In the USA there is
Organic Gardening and Farming,
Emmaus,
Pennsylvania 18099.

Rare Breeds Survival Trust,
The Ark,
Winkleigh,
Devon.
The RBST is 'an organisation devoted to the conservation, study, and promotion of Britain's lesser known breeds of domestic livestock'.

## Sheep

National Sheep Association,
Jenkins Lane,
St. Leonards,
Tring,
Hertfordshire.


Blackface Sheep Breeders Association,
24, Beresford Terrace,
Ayr,

The Bluefaced Leicester Sheep Breeders Association,
Crooklands,
Blencogo,
Wigton, Cumbria.

Black Welsh Mountain Sheep Breeders Association,
Brierley House,
Summer Lane,
Combe Down,
Bath, Avon.

Cotswold Sheep Society Ltd.,
The Old Mill,
Quenington,
Nr. Cirencester,
Glos.

Dartmoor Sheep Breeders Association,
Bilberryhill,
Buckfastleigh,
Devon.

Derbyshire Gritstone Sheep Breeders Society,
528, Red Lees Road,
Cliviger,
Nr. Burnley, Lancs.

Devon Closewool Sheep Breeders Society,
4, Cross Street,
Barnstaple,
North Devon.

Devon Longwoolled Sheep Breeders Society,
Hawkwill,
Sutcombe,
Holsworthy,
Devon.

Eppynt Hill and Beulah Speckled Face Sheep Society,
Market Street,
Builth Wells, Powys.

Exmoor Horn Sheep Breeders Society,
32, The Avenue,
Minehead,
Somerset.

Herdwick Sheep Breeders Association,
Glenholm,
Penrith Road,
Keswick, Cumbria.

Hill Radnor Flock Book Society,
Newmarket Chambers,
Abergavenny, Gwent.

Jacob Sheep Society,
Groves,
Jenkins Lane,
St. Leonards,
Tring, Herts.

Kerry Hill Flock Book Society,
Cantray,
Milford Road,
Newtown, Powys.

Leicester Longwool Sheep Breeders Association,
The Exchange,
Driffield,
East Yorkshire.

Lincoln Longwool Sheep Breeders Association,
Westminster Bank Chambers,
8, Guildhall Street,
Lincoln.

Llanwenog Sheep Society,
Bertheos,
Creuddyn Bridge,
Lampeter, Dyfed.

Lleyn Sheep Society,
Glasfryn Farm,
Chwilog,
Gwynedd

Lonk Sheep Breeders Association,
Jack Hey Lane Farm,
Cliviger,
Nr. Burnley, Lancs.

Oxford Down Sheep Breeders Association,
Boulton & Cooper Ltd.,
Malton,
North Yorkshire.

Rough Fell Sheep Breeders Association,
52, Milnthorpe Road,
Kendal,
Westmorland.

Shetland Flock Book Society,
Fairview,
Vidlin,
Shetland.

Shropshire Sheep Breeders Association & Flock Book Society
John Thornborrow & Co.,
II, Priory Terrace,
Leamington Spa,
Warwickshire.

South Devon Flock Book Association,
Bolitho,
Liskeard,
Cornwall.

South Wales Mountain Sheep Society,
13, Avon Close,
The Bryn,
Pontllanfraith,
Gwent.

The Teeswater Sheep Breeders Association Ltd.,
Edengate,
Warcop,
Appleby,
Westmorland.

Wensleydale Longwool Sheep Breeders Association,
27, Fountain Street,
Ulverston, Cumbria.

White Face Dartmoor Sheep Breeders Association,
Sawdye & Harris,
Newton Abbot,
Devon.

Wiltshire Horn Sheep Society,
The Homestead,
Kislingbury,
Northants.

## BACKYARD DAIRY BOOK
Andrew Singer and Len Street

## BACKYARD POULTRY BOOK
Andrew Singer

is book is written as propoganda and its aim
to provide enough basic information to
courage readers to begin home dairy produc-
on. Thousands have reduced their dependence
on factory food by growing their own, or
eping chickens. Backyard Dairying is a
rther step towards self-sufficiency.

ontents
1. Why start?
2. Basic Economics
3. Which animal
4. Making a start
5. Feeding
6. Milking
7. A cow
8. Dairy products
9. A little technical information
10. Cream, butter, cheese and yoghurt
    References and Bibliography

The author of the Backyard Dairy Book brings
the same treatment to the subject of domestic
poultry keeping. This book will appeal both to
the absolute beginner and the experienced
backyarder keen to try out new ideas.

Contents
*Stage One: Making the big Decision*
    Chapter One: Why keep chickens?

*Stage Two: Planning and Preparation*
    Chapter Two: A little technical background
    Chapter Three: Which system to use?
    Chapter Four: Which hens to buy?
    Chapter Five: What to feed them?
    Chapter Six: Confining and housing them

*Stage Three: Your Chicks Arrive*
    Chapter Seven: Rearing chicks into layers
    Chapter Eight: When to re-stock?
    Chapter Nine: The harvest—eggs, meat etc.
    Chapter Ten: Diseases and problems

*Stage Four: Other Poultry*
    Chapter Eleven: Ducks
    Chapter Twleve: Geese
    Chapter Thirteen: Turkeys
    Chapter Fourteen: Guinea Fowl, Bantams
    and Pigeons

The chapters dealing with goat management
e most convincing"—*Undercurrents*

A very useful book"—*Sunday Times*

t is remarkably clearly written and will be
elcomed by many questing clients in a prac-
tioner's waiting room"—*Veterinary Record*

"Andrew Singer's comprehensive manual deals
with all aspects of poultry keeping and is a
must for all potential poultry keepers before
making a start"—*Undercurrents*

# BACKYARD BEEKEEPING
Bill Scott

**Illustrated by Keith Spurgin**

A basic introduction to the gentle art which removes much of the mystique surrounding it. Part one explains the life of the hive and bee, part two describes the equipment and materials, and includes a design for a D-I-Y hive. Part three covers potential problems and their solution. The final section looks at the harvest and what can be done with it. Bill Scott is author of our very successful book, Food for Thought, and runs a wholefood shop in Truro, Cornwall.

"To be a successful beekeeper with a happy colony and plentiful harvest, you will have to invest time, effort and knowledge, and that's what this book can give you even if you are a complete novice".—*Here's Health*

"Unhesitatingly, I would say this book is a *must* for anyone contemplating keeping a hive or two of bees. Its simple clear style inspires confidence right from the start. The basics of beekeeping are explained in the clearest possible way and it even includes a section on making your own equipment."—*The Soil Association News*

## OTHER BACKYARD BOOKS

**Backyard Pig Farming**
Ann Williams

**Backyard Rabbit Farming**
Ann Williams

**Backyard Fish Farming**
Paul Bryant

**Backyard Farming**
Ann Williams